Become an automobile expert

A do it yourself guide to cars

1st edition

How to buy, inspect, maintain, troubleshoot and fix the most common problems in your vehicle

Theodore Ford

© Copyright 2018 – Alán Adrián Delfín Cota. All rights reserved.

The contents of this book may not be reproduced, duplicated or transmitted without direct written permission from the author.

Under no circumstances will any legal responsibility or blame be held against the publisher for any reparation, damages, or monetary loss due to the information herein, either directly or indirectly.

Legal Notice:

This book is copyright protected. This is only for personal use. You cannot amend, distribute, sell, use, quote or paraphrase any part of the content within this book without the consent of the author.

Disclaimer Notice:

Please note the information contained within this document is for educational and entertainment purposes only. Every attempt has been made to provide accurate, up to date and complete, reliable information. No warranties of any kind are expressed or implied. Readers acknowledge that the author is not engaging in the rendering of legal, financial, medical or professional advice. The content of this book has been derived from various sources. Please consult a licensed professional before attempting any techniques outlined in this book.

By reading this document, the reader agrees that under no circumstances is the author responsible for any losses, direct or indirect, which are incurred as a result of the use of information contained within this document, including, but not limited to, —errors, omissions, or inaccuracies.

Dedicatory

To all the ladies that had slapped me in the ass; with and without my consent.

CAR BASICS

Contents

INTRODUCTION 7
CAR BASICS - What You Should Know About Your Car! 16
CAR BASICS - The Components! 18
CAR BASICS-Tyres! 22
CAR BASICS-Air Filter! 39
CAR BASICS - Battery! 44
CAR BASICS - Car Hoses And Belts! 51
CAR BASICS -Automatic Transmission Fluid! 55
CAR BASICS- How To Maintain Car Engines For Fuel Efficiency! 60
CAR BASICS- The Radiator and Coolant! 67
CAR BASICS -Brake and Its fluid! 77
CAR BASICS -Power Steering Fluid! 84
CAR BASICS - Windshield Power Fluid!(*All you need to know*) .. 89
CAR BASICS –*Ignition!* 92
CAR BASICS - Steering And Suspension! 98
CAR BASICS –*Electrical!* 100
CAR BASICS- Avoiding Car Accident! 103
CAR BASICS -Emergency Road Services! 107
CAR BASICS- Road Safety Tips! 111
CAR BASICS - Emergency Items To Keep! 117
CAR BASICS -Preparing For Emergency! 124
CAR BASICS -Benefits Of Emergency Kits! 127

CAR BASICS -Advantages Of Emergency Road Services!..........130

CAR BASICS – Repair!..134

CAR BASICS- Auto Repair Tips!...139

CAR BASICS –GeneralMaintenance Tips!.................................141

CAR BASICS – *Washing!* 149

CAR BASICS -What You Should Keep In Your Car When Going On Trips!..153

CAR BASICS –*The Buying!*..157

CAR BASICS -Buying A Used Car!..161

CAR BASICS –The Insurance!..165

CAR BASICS –Warranty! 171

CAR BASICS -Buying From A Car Dealer!..................................176

CAR BASICS - Most Common Problems! 184

CAR BASICS -Essential Auto Tools!...187

CAR BASICS -Tools Should You Keep in Your Boot!..................189

CAR BASICS - Servicing!..194

CAR BASICS - Most Common Road Traffic Offences!...............199

CAR BASICS - Accessories Are a Must For Every Car!...............203

CAR BASICS– Specific Accessories For Specific Reason! 206

CONCLUSION 208

INTRODUCTION

Owning a car is a big responsibility. Whether you are picking out your very first vehicle or if you have owned one for years, it is important to take this responsibility seriously. A car in good condition is easier to drive, easier to take care of, and will last longer than one that has not been taken care of. Once you get the hang of it, taking care of your car is actually very easy. It only takes some basic know-how and a few helpful car care products to give your car the basic maintenance it needs to stay in great shape for many more years.

One of the best things you can do for yourself as a car owner is to familiarize yourself with the owner's manual that came with your vehicle. This little booklet has all the information you wll ever need to know about your car. It will also have lots of little hints that will help you stay on top of your vehicle's needs. Bookmark or highlight the portions that tell you what kind of oil your vehicle requires,the recommended tire size, and other important information. These are kind of like the vital statistics of your car; you will need to know these things to take care of many basic maintenance concerns, such as oil changes and tire rotation.

You can also help yourself stay on top of the recommended maintenance schedule by planning ahead for routine maintenance appointments. If you know that your oil needs changed once a year, simply plan ahead to do exactly that. Mark on your calendar the week that you will take your car in for its oil change and then make sure you follow through! If you know that this important appointment is coming up, you will have a greater chance of being able to make the time for it. It only takes a minute to pick a convenient week to do this and you will love how easy it is to simply take your vehicle to the nearest service station. There is no hassle when you already know what you have planned.

Another great thing you can do to help yourself care for your car is to create a basic car care kit. Your kit will contain all the basic necessities that your car needs to run smoothly and look its best. For instance, your kit might include car wax, leather conditioner, vinyl cleaner, and air freshener. Anytime you want to clean the interior of your car, all you need to do is grab your care kit and get to work. When you know where everything is located, you do not have to waste valuable time hunting for the bottles and rags you will need. Other items your kit might include are window cleaner, sponges, scrub brushes, carpet-safe spot remover, a handheld vacuum, chrome polish, and lint brush. Depending on what materials your car is made from, the contents of your kit may vary from what is listed here.

It doe not take a lot of money to put together a car care kit. Just about everything can be purchased inexpensively from an auto parts center or the automotive section of a department store. Once everything is collected, place everything into a handy bucket or bag. For safety's sake, read the labels carefully to make sure that you are following the storage instructions; also, keep all cleaning products out of the reach of pets and children.

Keeping your car in good condition is easy and it does not have to be expensive. A little planning is all it takes!

There are a few basic things you need to do to make sure your car keeps running. This is especially important if you are driving an older car. Even an older car can stay on the road longer and run better if you take care of it.

Some people do not care about checking the basic things and wonder why their vehicle refuses to run anymore. Some others do not really want to know and always have trouble which sometimes can be avoided. They just want to drive their cars and hope for the best, that is, it keeps running.

So what are the basic things that need checking? Checking the oil is very important to keep your car running as it should as mentioned briefly above. Making sure it is clean and at the level it should be. An engine thrives on good clean maintained oil. Dirty oil can make the engine run hotter which leads to other problems like making the coolant hotter or bursting a water hose. Never use inferior or cheap oil, this can really lead to problems.

Checking the oil is best done when the car has been sitting for awhile, so that all the oil has drained back into the sump, and you get a truer reading. Some folks truly do not know how to check the oil. So for those folk, on the side of the engine, or for a v engine such as V6 OR V8, look in the middle of the engine for a dipstick. It should be coloured on the top.

When you locate this, pull it upward and hold it horizontal so you can see where the oil level is. Wipe the oil off and replace the dipstick where it was. Wait ten seconds and pull the stick out again, as this is the level you really need to see. Holding the dipstick horizontal again, check that the oil level is between the full and add marks. If it is near the add mark and you have no oil handy get some as soon as possible, and fill it to the full mark by checking the dipstick as you fill it.

Another basic thing is the water coolant level which is in a plastic container to the side of the radiator. If you have a very old car it will be in the radiator itself. If you have a newer car with a plastic container the coolant should be green and be between the minimum and maximum level marks on the container.

When there is a need to add some more coolant, never take the top off the plastic container or radiator straight after the engine is turned off. The water coolant will shoot up and burn you as it is still under pressure and hot. All the coolant will also come out and will have to be replaced.

The third basic thing to check is your tyres regularly. Maybe every other week. Tyres running on the correct pressure give you a better ride, better fuel economy and a better stable feel to the steering. Look in your owners manual for the correct pressure for your tyres.

The fourth basic check is only for those with power steering. There is a small usually black metal container with a metal or plastic top on it. This is the power steering oil reserviour. To check the power steering oil turn the plastic or metal top anti-clockwise to remove.

The removed top should have a short measuring plastic length under it, which goes into the power steering oil. Check to see the oil on this length of plastic or metal is between the add and full. Top it up if need be with the right power steering oil for your car.

If there was no oil on the length of plastic or metal then it needs oil fast. Sometimes this is never checked. One way to tell you the power steering oil is low is there will be a wining noise, when the steering wheel is turned in any direction.

The fifth basic check that can be done is oil in the transmission. Many folks never check this themselves and it can lead to costly repair bills especially in older vehicles. This is best checked when the engine is running. So you can check it when you have come back from somewhere and the engine is still warm.

Make sure that the car is in park with the hand brake on to check this. Open the bonnet and look near the back of the engine for another larger dipstick. While the engine is running keep your hands clear of the fan in front of the engine, which is rotating fast. Locate the dipstick and pull it out holding it horizontal to get a reading. Then, as before do it a second time after you wipe it dry with a rag or paper. The oil should again be between the add and full mark as with the motor oil.

Unfortunately we live in a world of unscrupulous people who work on your car, and something as simple as no oil in the transmission, can cost more than it should. All that may have needed doing was topping up the oil. An example of this which was experienced by someone was the second gear on his older car was changing roughly. When checking the transmission oil he found it was low and topped it up. When driving the car again the problem had gone away and second gear was changing smoothly again.

A lot of people do not know how to do these basic checks, so I hope this will help many people. Not doing these checks regularly can add more to your motoring bill.

A car will always run better and longer if you give it these regular basic checks. You also avoid costly repairs in less often circumstances. If you can not afford to replace a car as often as you like, then these regular checks are of great benefit to you and your car. Always remember if you look after your car, your car will look after you.

CAR BASICS - What You Should Know About Your Car!

Most people do not know much about their car. They do not want to trouble themselves with that, they think it is not important at all or simply think that pushing the pedals and turning the steering wheel is enough trouble already. But in situations (like an arm-chair conversation about cars) people find themselves sticking out like a sore thumb.

That is the more harmless part. The lack of car-knowledge could be potentially troublesome. What happens in case of a needed car repair and you do not know squat about your car or even what to repair?

What happens when you want to sell your car and do not know the basics? The chances of a proper sale is seriously reduced.

So for people who want to know more here is some useful things you should know about your car:

- Car manufacturer

- Model and type
- Type of engine (petrol or diesel)
- Engine configuration ("V" or in-line)
- Type of transmission (manual or automatic and number of gears)
- The correct type of engine fuel your car use
- The vintage of your car (the manufacturing year)
- The correct mileage or kilometers on your car
- Engine size (in cubic inches or centimeters)
- Basic orientation under the hood (in case of repair)

Seems overwhelming? It is not at all. You can squeeze all of this into one sentence and it will take only a few minutes to learn this. Who knows, maybe it can get you more interested in cars and perhaps, one day, do some car repair yourself.

So, if you are a car driver, this is some facts or knowledge you have to have. Not having it is like sitting at the table not knowing the purpose of the fork and spoon.

CAR BASICS - The Components!

This is the pre level in automotive repair before proceeding to the next level of repairing a car. This pre level is about knowing the car components and the use of those components. This is the most basic level and the most important part to know before touching any tools in automotive repair. This basic components is divided into 2 main part.*1.Firing component . 2. Fuel component.* And they shall be discussed accordingly.

A. The Firing Component

➢ *Spark Plug*

The use of spark plug is to bring sparks in the firing system of the engine before the combination of fuel to make the engine start.

➢ *Battery*

To start the engine a car must have a battery just like a remote control items which have this component too.

- **Starter motor**

This component is a mechanism to make the gearbox flywheel spin order to turn on the engine.

- **Firing distributor**

In order to distribute the sparks, engine must have this as well as the spark plug cable before the engine can run. There 2 types of firing distributor which are electronic and non electronic firing distributor.

- **Plug cable**

It is an important part to make the sparks distributed to all the spark plug and the engine can run on their own consistently.

- **Ignition coil**

This is a mechanism to keep the sparks energy and it goes along with distributor to spread the sparks all over the engine using a plug cable.

B.The Fuel Component

Now we move to the fuel system of a car. It is very important to know that fuel is a main component also in engine system,without fuel the engine cannot even run and the rest of engine components also cannot play their role.

> ➢ *Fuel pump*

To spread the fuel to carburator or injector we need this component to suck fuel from the fuel tank.

> ➢ *Fuel filter*

To filter fuel from dust or any tiny items from fuel tank after reaching a fuel pump.

> ➢ *Fuel tank*

It is tank to keep the fuel and always located at the bottom of a car.

➢ *Fuel injection*

As a new technology in automotive fuel system after carburator this components act as an injector to maintain the ratio of air and fuel in keeping the idle system of the engine to run.

➢ *Carburator*

The old fuel system now being replaced by fuel injector and the role still the same.

CAR BASICS-Tyres!

Tyres are essential parts of a car or vehicle.Its role and essentialities cannot be over-emphasized because without tyres movement is not possible. Safety of the users and its durability largely depends on its quality.It is expedient that quality tyres are used for a better performance of car.

Maintenance Of Tyre

Keeping a check on your tyres are very important for optimal performance. Checking pressure regularly and making sure the correct pressure is maintained is crucial. Low pressure will slow down the car and too much causes wear and tear to the tyres. Pressure is best checked when the tyres are cold and run for less than a mile.

Another issue is the lack of balance. Badly balanced wheels and breaks will cause wear on the middle of the treads. This is extremely unhealthy for the overall car. Check your tyre treads and the depth of your tyres regularly. If the depth is between 1.6 - 3mm then it is about time you get new ones. Checking the depth of the tyre is best done by a professional.

1.Balancing

A common symptom of unbalanced wheels is an uncomfortable driving experience due to vibration of the vehicle and or steering wheel. Wheels that are out of balance can also cause damage to your car and cause your tyres to wear much faster than they should - costing you money in repairs that could be avoided with a simple re-balance.

Usually the first sign of out of balance wheels is a wobbly steering wheel when you are driving above a certain speed - normally this is at high speed on a dual carriageway or motorway at speeds above 50 mph. Wobbling of the steering wheel often suggests that your front wheels are unbalanced. If you are experiencing vibrations in the seat or back of the car, it generally suggests that the wheel imbalance is in the rear wheels.

Sometimes, however, these vibrations are dampened by the weight of the vehicle - masking that there is a problem. Therefore you should get your wheel balance checked regularly, preferably at a service or when you are having your tyres replaced.

Incorrectly balanced wheels can cause damage and premature wear to:

Suspension

Tyres

Steering components

Rotating parts

The benefits of correctly balanced wheels include:

Better handling

Safer driving experience

More comfortable, smoother ride

Longer-lasting tyres

Fewer vibrations

Better fuel-efficiency

Wheels are balanced on a special machine which rotates the tyre and wheel to calculate the correct balance required. If there is a problem, one side will be heavier than the other, and the technician will then balance the tyre by applying a counter-weight on the opposite side. When a tyre is imbalanced, it will rotate asymmetrically, causing wobbles and ride disturbances which will increase with speed. This can in turn knock out your suspension and cause more costly problems.

Sometimes the problem is not with wheel balance but with wheel alignment, which can be maladjusted due to aggressive driving, hitting curbs or driving over potholes or speed bumps. This is another simple fix at most garages or tyre specialists, and returns the wheels to the specifications given by the car manufacturer. Wheels that are out of alignment can also affect the quality of your car's performance, handling and the lifespan of your tyres, so checking both alignment and balance is important if you want to save money on repairs, tyre replacements and fuel.

2. Rotation

It is surprising to think how much maintenance is necessary for your tyres and wheels. If you think about it, we are a society that drives around to every single place whether it is close by or not. It is no surprise to see how damaged tyres can get.

Imagine if you are the type of person that only turns left. More force will be placed on your left tyres because of the same movement occurring over and over again. The damage that occurs on your tyre will have an uneven drive for you because of how much wear has taken place to the one tyre and not the other. This is an uncomfortable way of driving and could eventually damage the mechanical systems that allow the car to drive straight. The wear on the left tyre could cause you to lose control of the car depending how bad the damage is. Due to the nature of rubber punctures are likely to occur. It is a well-known fact that a punctured tyre is harder to control than a bit of wear.

With that in mind, tyre rotation is the process of allowing even wear on all tyres. By moving from one wheel to the other the mechanic can get the exact amount needed for a comfortable drive. There are many causes for uneven wear that can be avoided.

Make sure that your wheels are properly aligned. Unaligned wheels put more force on one side than the other. In the end you will have a lopsided car prone to accidents and wheel damage. It is important to get the wheels aligned as often as needed.

Over inflation is another cause of uneven wear. Under inflation can also cause uneven wear on the outside of the tyres.

Other mechanical elements could also be the main cause of this kind of tyre damage. If the shocks do not allow absorption of the force your tyre could suffer from pressure applied by the car itself.

Drifting around a corner is one of the harshest procedures that can damage the tyres so much that you might have to replace them the same day. Involving hand brake turns and acceleration at the same time will give you the worst uneven wear you will ever see.

To get this problem sorted out, you have to see a tyre specialist who can rotate the tyres. It is important that regular tyre maintenance is done for a safer drive.

3.Alignment

It is always a good idea to get new car tyres checked for alignment. It is quite easy for tyres to get out of alignment; you only have to hit a bad pothole in the road for the suspension to get bumped out of alignment. What does it mean exactly for car tyres to be out of alignment? It means that they are not quite straight - and so there will be a great deal more wear and tear on them. Having just paid a sizeable sum of money to have them fitted, you hardly want them to start wearing out unnecessarily.

How can you tell if tyres are not properly aligned? Very often the vehicle will start to shudder at certain speeds, or the steering wheel might vibrate at higher speed. If the problem is not corrected you will soon see uneven wear on the tyre treads as it means that one or more tyres are being dragged to some extent rather than running straight and true. There are other problems caused from not having your tyres properly aligned.

* **_Higher fuel consumption._** When tyres are not running straight there is drag which means the engine must work harder to make the car go. This relates to more fuel consumption.

* **_The car feels as if it is wandering_** - the steering wheel may pull to the left or right.

* ***The ride will not be as smooth*** as one or more tyres are constantly being dragged to correct their direction.

* ***Handling and control will not be as efficient and can cause accidents.*** It can also cause driver fatigue.

When you get a wheel alignment done it is necessary to have all four tyres aligned at the same time. Only then can you be sure that they will all be travelling in the same central direction. This should be done at least once a year or every 20,000 kilometres or so. It may need to be done more often if you constantly drive on bad roads that contain lots of potholes, or if the car has been in a minor prang.

To check if your vehicle needs an alignment, choose an empty road and while driving slowly, let go of the steering wheel. If the vehicle pulls to the left or right, or if the steering wheel rotates without you touching it, you can be pretty sure that the alignment is out. At higher speeds take note if the car shudders or if the steering wheel seems to vibrate. See if it seems hard to turn a corner or do parallel parking. Such problems can often be caused by your wheels being out of alignment - or out of balance.

Having regular maintenance such as a suspension check on the vehicle often means that other problems can be picked up before they translate into something serious and more costly. And since it makes the vehicle safer, easier to drive and saves you running costs, getting it done should certainly be something that you would not think twice about.

Checking The Tyre Pressure

It is recommended that before long journeys, and also on a regular basis, that you check your tyre pressures, as having the pressure too low causes excessive wear and tear to the edges of the car tyres, which will shorten their life. Not only this, it also uses more fuel, so ensuring the optimum tyre pressure will save you money. Keep in mind however that you should be careful not to over inflate tyres either as this causes unpredictable handling.

When checking tyre pressure make sure the tyres are cold, and do not check after a drive as this will warm the tyre and give an incorrect reading. Remember to take into account whether the car is heavily laden which could affect the reading.

How To Change A Car Tyre

There is a every tendency that everyone will experience a puncture at sometime in their driving lives. But it is surprising how many people still do not know how to change a car tyre and rely on the help of a passer-by.

In all other situations changing a car tyre is not as difficult as it may seem. This book is a guide to how it is done:

1. Find a safe place to stop, on flat ground. If it is dark look for somewhere with adequate lighting. Park as far away from traffic as possible and put your hazard lights on.

2. Turn the engine off, pull the handbrake on and put the car into first gear or 'park' if it's an automatic.

3. Find the tools for the job. These will include a jack, wheel wrench and the spare tyre. All are usually located in the boot of a car.

4. Remove the hubcap or plastic wheel cover using the flat end of the wheel wrench or a screw driver.

5. Loosen the wheel nuts by turning half a turn in an anti clockwise direction. Do not remove them. That happens at a later stage.

Tight wheel nuts can be the hardest part of changing a car tyre. If the wheel nuts are proving stubborn use a metal tube to extend the wheel wrench's handle and provide extra leverage. Alternatively place one foot on the handle of the wheel wrench and carefully use your body weight until the nut is loosened.

Many cars have a locking wheel nut to prevent theft. This requires a special attachment which is often found in a car's glove compartment.

6.Look for the jack and take it out.Look for a reinforced lip on the underside of the vehicle close to the wheel arch.

When the jack is securely attached raise the car until there is enough room to remove the punctured tyre and replace it with the fully inflated spare.

7.Now continue to remove all of the wheel nuts and place them somewhere safe nearby- not near an open drain! Remove the punctured car tyre and replace with the spare tyre.

8. Replace and tighten the wheel nuts. Do not try too hard to tighten them until the car is off the jack.

9. Slowly lower the car down off the jack. Now tighten the nuts again. Remember to return the jack and the wheel wrench to the boot of the car.

10. Replace the hubcap or plastic wheel cover and continue with your journey.

How To Buy Tyres

In order to be able to buy car tyres one must familiarize themselves with some basic information about tyres, and how to choose the proper tyres for your car. Although many tyre shops can give you proper advice on tyres and which will be best for you, it is still important for you yourself to know about tyres, in order to save time, to go in knowing what you want, and to avoid getting lied to when you do buy the tyres.

> ***Basics of How to Buy Car Tyres***

One of the first things to distinguish different tyres is their size, and type.The dimensions of tyres are usually written on the sidewall of every automobiles tyre. An example of this code will be "P225/60/16 95H M+S." One of the most important things in replacing tyres is finding the correct size to buy. The tyre size is indicated by the3 sets of numbers indicating its width, aspect ratio of sidewall compared to width, and diameter of the rim.

- *Type, Speed Rating and Load Index of Car Tyres*

The type of tyre is indicated sometimes by either a P or LT meaning passenger (P) or a light-truck (LT) tyre. The next set of alphanumeric code (e.g 95H) is the load rating and speed rating. These numbers and letters indicate the tyres ability to dissipate heat at high speeds, and the load carrying capacity. The final set of number indicates the purpose of tyre. For example M+S is all season tyres. These basics are very important to know when buying car tyres.

In order to get the best performance and fuel economy is recommended to use the manufacturers recommended tyre size and dimensions. The manufacturer of your vehicle will usually have this number on a sticker in your inside panel of the driver door or the gas tank door.

- ***Buy Car Tyres Based On Treadwear***

To determine the wear and tear time you should look for the tyres UTQG or Uniform tire Quality Grading. This will tell you the tread wear, the temperature resistance, and traction of the tyre. The UTQG rating of wear and tear is a numerical rating system and an alphabetical grading system like A to C is for temperature, and AA to C is for traction.

- ***You Can Buy Car Tyres That Are Technologically Advanced***

Many manufacturers have technologically advanced tyres that make your driving experience more comfortable, and slightly safer, consider these when you buy car tyres for more of an edge to your driving experience. One of the most popular types of technologically advanced tyres is run-flat tyres. Run-flat tyres are designed tough to withstand punctures, and flattening out when a nail or tear affects it. These tyres have a tough sidewall and it keeps it from deflecting. Some run flats have different ranges you can drive on them while flat; one of the best systems is the PAX system which can take you up to 125 miles at 55mph.

- ***Read Reviews Before You Buy Car Tyres***

Before you purchase any car tyres, you should look up the brand and model, including all the extra technology that it comes with and read reviews about it. Personal experience can really give you better understanding about what others think about this, and also professional reviews can sometimes be better than any of the research and studying you can do on your own.

CAR BASICS-Air Filter!

Its Functions And Maintenance

A car cabin air filter is a very important component of your vehicle. It is a filtration device attached to the air intake of car's ventilation system. The main function of this device is to improve the quality of air penetrating inside your car as well as of the air already present inside the car. Although most of the filters are made using pleated paper system, a host of other filtration media is also present. An intelligent blend of paper and cotton is perfect for the filter. However, other models using miniature paper filters are also available.

This filter is totally different from the internal combustion filter that sits under the hood of the car and prevents the dust from getting into the car's engine to facilitate smooth functioning. Distinction between the two is very clear and having its knowledge is very important. Both of them hold an important place in your car's smooth functioning and hence you must check their quality while buying.

The cabin air filter is responsible for the kind of air you breathe in while you are sitting inside the car. Installing a high quality filter is the only option if you want to offer fresh and pure air to those who are sitting with you in your car. The green filters are the best options for the people who are more worried about the allergens. However, every air filter needs to be replaced after a period of time to ensure quality of air is maintained inside the vehicle.

How To Change A Dirty Car Air Filter

Changing them regularly will help you keep your car working like brand new without any issues. It will also help you save fuel since good airflow will keep it healthy.Your car needs to breathe fresh air similarly like we need clean oxygen for respiration.

Paper air-filters keep the cars motor free from dust and insects. Changing or clean your car's air channel at the proper time to keep fresh air flowing to your car's drivers in is essential to keep your car's performance optimal and at its best.

They are really cheap to buy and easy to change, so you can do this yourself.

Here are the steps required to change it:

• Obtain the exact replacement air-filter that fits its retainer cup. Make sure that you have the correct one.

• Take proper guidance from your owner's manual or ask a local mechanic.

• Park your car on a ground level and keep it on the parking breaks.

• Open your car's bonnet (hood) and secure it tightly.

• Locate the unit housing it is kept in, which is generally on the top of the vehicle's engine chassis.

• Carefully remove the top of its cover.

• Loosen its hose clamp which acts a sealant for the air conduit.

• Unscrew all tightened screws securing the cover.

• Keep the screws and removed parts together in a tray so you can find them later and assemble systematically.

• Pull the safety cover slowly out of the air conduit to lift it up.

• Now the bottom part of the housing will be visible and the dirty one will be at its bottom; take it out and discard it.

• Clean the dirt out of the housing.

• Replace the new part and replace the covers that you unscrewed. You have now successfully installed a new one for your car's engine.

Doing this regularly will ensure two things for you; one, peace of mind that the car is breathing enough oxygen to help the petrol ignite properly and efficiently to give you maximum mileage and two, that the engine is cool enough to run for longer distances.

If your car gets really hot even after running for a really short distance it means only two things, either the coolant is leaking and needs change or the air filter is so clogged up with dust and other grime that it is impossible for the engine to breathe which leads to heat buildup.

Always check the air filter at regular intervals to keep your car breathing properly and enabling it to run with maximum efficiency. Keeping your car running like new, every time!

CAR BASICS - Battery!

Maintenance Of Car Battery

It is very important to understand how your car works. The car battery is very important, as without it, your car cannot start. Many things in your car like the clock, mobile charger and radio, work with the help of your energy box even when your car is turned off. No battery can last forever; however, proper maintenance can give it an extended life.

Every car has its battery placed under the hood and is easy to spot it. Most car energy cells are large with either rectangular or square shaped boxes connected with two cables. These cables should be kept clean and corrosion free. If you spot small crystals or white powder on them make sure you clean them immediately with the help of a wire brush.

Generally a battery which is a 12 volt type will be made up of two 6 volt cells having positively and negatively charged lead plates and separators that are insulated. The solution inside the cell is an electrolyte which consists of two thirds distilled water and 1 third sulphuric acid. The electrolyte solution and the lead plates interact to produce a chemical energy, which in turn is converted into electric energy to produce electricity for your car.

When handling a batteries one should be very careful and should use proper covering and clothing, gloves and goggles. A torch should be used instead of a match when working near the battery, switch off all possible electric appliances. Your car battery is likely to produce gases which are highly explosive.

Depending on your driving ways and maintenance of your car battery, generally a car battery is replaced with a new one after every three to five years. The way you start and stop the car engine or the climatic changes also play a very important factor to determine the life span of your car battery.

When your car is facing a starting problem, your car may need jump starting. Jumper cables are attached to a working battery and then to your car battery. The positive cable should be attached to the positive terminal of the dead energy box and then to the positive terminal of the working car battery. Now, the negative end should be attached to the negative terminal of the working car battery and the other end to a metallic part of your car engine. Make sure you do not attach the negative end of the cable to your dead battery under any circumstances. Now you can start your car engine with the help of the working energy cell. Let the engine idle for a while until you switch on the headlights of your car with your energy cell. When you turn on the lights of your car you can save the car computer from any voltage fluctuation or sparks. Once you have done all this, you can start your car with the help of your dead battery and remove all cables in reverse order. Make sure the clamps do not touch each other when removing them.

Learning simple methods to take care of your batteries will save you from problems and will ensure your car battery has a longer life.

Time To Replace Car Battery

Every mechanical innovation requires an energy source that will offer power for its movement. In the case of automobiles, this source of power is battery. The batteries supply power to every part of the car that requires electricity to work, like the ignition system, air conditioning, stereo, and headlights. So, you can understand very easily what will happen to your car when the battery dries out or becomes out-of-order.

Most cars of today include GPS systems and various other high-tech features, which works with the help of a battery. Thus, when your automobile battery becomes faulty or dead, none of these high-tech features function. Hence, if you do not want to face such situation, you must know all the signs that tell your car battery is about to die. Like all the other parts of your car, batteries also have a specific lifespan; you need to replace them within that lifespan to avoid these troublesome situations.

Refurbished Batteries an Alternative Option

Replacing car batteries does not mean that you have to replace it with the new ones. You can also use refurbished batteries as a replacement. These are quite easily available and also cost less than the new ones. Best part about these refurbished batteries is that they come with great performance.

Few Symptoms of Failure

To understand whether the battery is performing well, you must know their testing methods. Use a voltmeter to know the voltage output, without and with a load. If the voltage drops below 12 Volt without any load, be sure that your car battery needs an urgent replacement.

Apart from this, the cranking sound of your car and difficulty while starting the vehicle, are some symptoms that say your vehicle battery need an urgent replacement.

The Perfect Time

There is no such perfect time of replacement. This mostly depends upon the maintenance, amount of usage, climatic condition and quality of the battery. If any of the above-mentioned features go wrong, be sure that it is time for a quick replacement of your car battery. If you consult a mechanic, he will also recommend you to change the battery once in every three to four-years.

You can replace the faulty battery with reconditioned batteries, but before that, make sure it comes with a warranty. This is because the warranty means that your battery is of good quality and will run for a long time.

Apart from all the above, the best method, of determining the replacement time of a car battery, is through inspection. Consider all the symptoms and then decide when you can replace the battery.

CAR BASICS - Car Hoses And Belts!

Car hoses and belts are the most essential components of the car and it is imperative that you take special care of them. Any problem with the hoses and belts of your car would mean that the engine would face malfunction and deteriorate the efficiency of the car. Hence, you should conduct monthly inspection of your hoses and belts with the help of a trained technician to ensure a long and healthy life of your vehicle.

Hoses

Like your belts, hoses are equally important and are used in many accessory components of your car. Maintenance is essential from time to time. Hoses usually face leaks and this can temporarily be fixed using duct tape wrapped around the hose, but it is recommended you get it replaced by a new one. When a hose feels hard or makes a crunching sound when squeezed, it means that it is deteriorating and should be replaced with a new hose. This applies for both car hoses and belts that you should have them replaced as soon as possible before they start causing problems to the other components of the car.

Both car hoses and belts are available in the market and are also relatively cheaper than the other parts of the car. These are extremely vital components of the car and should not be ignored. Regular checks on them and speedy replacements are essential.

Things to Remember

- Inspect hoses regularly and get it checked by a mechanic for proper maintenance
- Squeeze the radiator hose to check whether it is in proper condition
- Regularly examine the belts
- Check the tension in the belts and make sure it is not too stressed.

Belts

The belts are essential parts of the car and are needed to efficiently operate the accessory systems in the engine. These include the alternator, air conditioner compressor, power steering pump, and the water pump. Without fully functional belts, these parts will not be able to perform adequately. Car hoses and belts work in the background but are equally important for the correct functioning of the car.

You should regularly check your car hoses and belts. Belts with any signs of cracking, missing pieces or fraying need to be replaced immediately before they cause further damage to any of the other parts they are associated with.

Also make sure that the belt tension is properly checked and adjusted from time to time. If the tension between the belts is too tight, they will deteriorate sooner and will cause the accessory components to malfunction. Bottom line is that you should regularly have your car hoses and belts inspected.

CAR BASICS -Automatic Transmission Fluid!
How it works

The fluid does what its name says it does, it makes sure that there is silky operation with the car's transmission. It also cools the working parts of the car and serves as a way to transmit energy from the engine to the transmission.

It depends on what type of car transmission you have. It can be automatic or manual. More or less it has the same purpose - to make your car work smoothly.

One of the most commonly over looked aspects of a vehicle is the transmission. Many people tend to either ignore their transmission all together, or simply do not understand how it works. And that is fine and dandy until the day comes that things start to get ugly with it. A busted transmission is another way of saying a car that is not going to be running any time soon without some serious work. The transmission is the power source, and it provides proper application of the power of your vehicle. So if you want to ensure your vehicle's transmission lasts just as long as possible (and most will last the entire vehicle's life time if maintained properly), then you will want to get on top of changing your automatic transmission fluid.

Inspection

Check automatic transmission fluid every month and whenever the transmission goes wrong.

1. Park your car on a straight land and start the engine, leaving the gear in neutral. Wait for the engine to warm. If the car manual does not say anything, let the engine go during this procedure.

2. Find the rod of the automatic transmission fluid, located behind the engine. Transmission rod is usually shorter than the rod motor oil, but it looks like the rest. If you are lucky, it will be able to find it with a label on.

3. Pull the rod and remove it entirely. It could be very long.

4. Wipe the rod with a cloth, put it back in the engine, push it all the way and remove it again.

5. Look at the rod tip. Look if there are two different markings: one for reading cool and one for hot. If so, read the one for hot. If the fluid does not reach the line marked "full", add liquid.

6. Add liquid to the hole through which the rod got out. Use a funnel with a long neck and narrow. Add a little bit and check the rod after each cast. It is easy to add transmission fluid but it is difficult to remove if you put too much.

7. Put the rod in place inside when finished.

Tips

- Consult your car manual to see what kind of transmission fluid should you use.
- On some cars, the engine should not be running while checking the fluid, so be sure to consult the car manual.

- Automatic transmission fluid should be consumed, so if you have a low level, means that there is a leak. Do not ignore leaks and not run with little transmission fluid, can lead to expensive repairs.

Lifespan And Replacement

In the past, automobile manufacturers asserted that you could drive for 100,000 miles without needing to change the liquid. Mechanics have long maintained that to be false. Recently, manufacturers have backed down to a more reasonable 30,000 miles. Now, if nothing goes wrong, you should be able to drive for much longer than 30,000 miles without problems. But, by the time you notice a problem, you need to top off your automatic transmission fluid(ATF) immediately.

The process for a trans flush is messy and complicated. You must first remove the transmission pan and replace the filter. These are going to be in different places for every vehicle. However, that will only solve about half of your problem. Much of the fouled-up ATF will remain in the transfer case, clutch drum, trans cooler line, torque converter, valve body, and other places.

If you are well-versed in the workings of your auto, this might not be a daunting task. You will need to find a level spot such as a driveway or parking lot. Then, you need to drive your car onto blocks or jack it up with jack stands. Be sure to place chock blocks behind your rear tires so the automobile can not roll. You will then need to drain the transmission fluids, bolt everything back to where it goes, and fill your car up with new liquid.

If this all seems like too much work, your best bet will be to call a local mechanic for a trans flush. Whatever you decide, this is not a problem you can ignore.

CAR BASICS- How To Maintain Car Engines For Fuel Efficiency!

The Engine

The car engine is certainly the main part that determines the usefulness of a car. No matter how luxurious a car you have bought, unless you have the right engine, you will not be satisfied with its performance. However, to ensure that you have purchased the right engine, you have to understand the basic features of the car engines. You have to understand the basic parts of the car engine. This will help you understand the way they function and at the same time, you will be able to identify the right engine for your car.

Well, one of the most important parts of an engine is the cylinder. There are different types of cylinders in a car engine and they work differently. This is why different manufacturers use different types of cylinders that make their cars suitable to a specific type of terrain. So, before you choose the right car, you have to ensure that the number of cylinders as well as the pattern of them is suitable for that. This is one very crucial factor that decides the performance of a car.

The next important part of an engine is the valve. This is also a very important part since it controls the fuel and air that will be released to the piston chamber. This is countered by the exhaust valve that releases the burnt fuel from the piston chamber. Eventually, this is released through the tail pipe.

Now, we have the piston of the engine to consider. Basically, this is the power device in your car engine. The combustion stroke forces it down while the flywheel and crankshaft bring it up with the momentum of the exhaust stroke. Actually, the flywheel is designed as a varying weight on crankshaft and it stores the momentum. The crankshaft keeps on rotating and it receives the movement from the stroke of the piston. Thus, the crankshaft becomes the initial drive line of the car.

Because of so many mechanisms going on, the car engines produce a lot of heat. This is why the cooling systems are necessary to keep it working for a long time. For that, the antifreeze is stored in the radiator to cool down the block of your Car engine parts. This is done by the radiator fan. This antifreeze is basically, a constantly moving path of hoses and compartment to cool the engine block.

However, for an engine to start, it needs a spark. This is provided by the starter system that generates the spark to light the fuel air mix compressed in the piston. This system is run by battery and it also lends movement to the crankshaft. The lubrication system in the engine makes sure that all these parts are moving smoothly and without damaging frictions.

Fuel efficiency of car engines is considered as an imperative and essential factor when buying a new car, all thanks to the high fuel prices. Most car makers have taken action to provide more fuel efficient cars that return more than 80mpg of economy however, there are millions of old cars that people use for their daily commute, by following the simple guidelines below, they can also achieve maximum fuel efficiency and the lowest carbon emissions from their car engines.

- ***Keep the Engine Appropriately Tuned***

It is very important to follow the maintenance schedule of the engine, for instance, replacement of air filter, spark plug and oxygen sensors, also keep an eye on any on-board diagnostics malfunctions in the Engine Control Module. The most important is to change the engine oil as instructed in your car's manual.

- ***Fuel Evaporation***

Fuel evaporation is another parameter that needs to be attended by tightly closing the fuel tank lids and by parking in shades, believe it or not, can improve your mileage by as much as 10 percent.

- ***Use the Recommended Grade Engine Oil***

The most important factor to keep your car engine at fuel efficient state is proper engine oil with low-kinematic thickness that is also referred to as low "weight" engine oil. You can get better mileage by 1-2 percent by using the recommended grade of vehicle engine oil. For instance, using 10W-30 motor oil in an engine intended to use 5W-30 can worsen your fuel efficiency by 1-2 percent. Contemporary engines have such accurate tolerance that very trivial oil is often required, thicker oil, such as 10W30 or 10W40, may not lubricate as well, because it will not pour as swiftly into key oil ways and fractures. You must check the owner's manual for suggested viscosity and ask for it exclusively when oil is changed.

- ***Minimise the Idling***

Your engine wastes fuel when car is in idle state, simply because you are burning fuel while going nowhere. If you are waiting for someone and you know that it will take more than 30 seconds, switch your engine off because you are just burning money.

- ***Keep Tyres Appropriately Inflated***

You can get up to 3.3 percent better mileage by keeping your car tyres inflated to the suitable pressure, this reduces the resistance so less power is required to move the car. Under-inflated tyres can lower your engine efficiency through more fuel consumption. The correct tyre pressure for your vehicle is generally found on a sticky label in the driver's side door jamb and or in your owner's manual. Do not use the utmost pressure printed on the tyre's sidewall.

- ***Smooth Driving***

The way you drive your car also plays a huge role in fuel consumption, by driving fast, you might save 5-10 minutes, however by smoothing down you can improve the fuel consumption of your car's engine, if you normally drive on a motorway at 70mph, try changing it to 60mph can increase the fuel economy by up to 4 miles per gallon. Avoid as much as you can from putting your foot down on the throttle and from stomping on the brake paddle, this will not only save you on fuel costs but also money on wear and tear of brakes.

CAR BASICS- The Radiator and Coolant!

Radiator

When you drive, your car's engine produces a considerable amount of heat. And it is the radiator's job to eliminate this excess heat so that your vehicle can operate smoothly and safely. It is the main component to your vehicle's cooling system, and without it, your engine would overheat and cause damage to other auto parts every time you drive.

Coolant

The radiator does not manage the heat levels in your engine all alone; the heat exchange process is coupled with the utilization of a well-known, heat-absorbing liquid called coolant. Coolant must be replaced in your radiator on a routine basis in order to maintain a properly-functioning radiator and engine. Coolant is also known as anti-freeze, because it also prevents the engine from freezing up in cold weather.

How They Works

A radiator is typically made from aluminum because it is a terrific heat-dissipating metal and low in weight. But it can be made from steel and other metals as well. Radiators work by sending coolant through the inner components of the engine to absorb heat; and once enough heat is absorbed, the coolant travels back to the radiator to be cooled down, and the cycle continues as you drive.

Car Radiator Leak Repair

Radiator is an important part of an engine of a car, performing as a heat exchanger. The coolant inside it absorbs the heat from the engine block and passes this to auto part where the belt-driven or electrical fans cool it down. Now, this radiation system may develop leakage with time and usage. Being a significant part, it calls for instant fixation. This fixation is formed of a few easy and basic steps, which can be done without the help of any mechanic too.

Firstly, the spot from where the leakage is occurring is to be located. Check whether there have been any puddle deposits and or holes in the radiator. This can be done by washing the radiator and the hose with water and then searching for the signs of any leakage. If you are unsuccessful this way, the radiator can be removed altogether to find the spot.

Once the place of leak has been detected, it should be patched up. A duct tape usually serves the purpose if the leakage is in the hose pipe. However, a car radiator leak sealant is to be used for leakage detected in the radiator itself. The proper instructions should be available with the kit for carrying this part out. Pepper powder acts as a good sealant in this case as pepper tends to swell up and thus patching up the leaking area totally.

Once the leak is sealed off completely, correct amount of coolant is to be added. Now, close the radiator cap and cross check for any more leakage. If no leaks are found, the car can be used again normally. However, a regular examination of the radiator should spare you all these trouble!

How To Maintain And Replace Your Car's Radiator

Most of the fuel energy in a car is converted into heat (approximately 70%), and it is the job of a car's cooling system to take care of that heat. A car's engine has numerous parts moving constantly to generate power. These moving parts create friction that results in high temperatures. Though motor oil is pumped throughout the engine to provide lubrication, it is not adequate to surmount all the excessive heat. Consequentially, several parts of the engine reach temperatures high enough to cause damage. This is where a cooling system becomes necessary in a car.

The engine must be kept moderately cool to operate normally and to avoid seizures. To prevent this from happening, water and coolant liquid is pumped through several components of the engine to absorb heat. When the heated liquid exits the engine, it re-enters the radiator, where drawn through a number of internal folds and chambers, it is cooled. Designed to transfer heat from the mixture of water and the coolant coming from the engine, the radiator is a type of heat exchanger. The radiator is helped by a fan that blows coolers outside air to speed up the cooling process.

Most modern vehicles use radiators made of thin aluminum tubes. Several tubes are arranged parallel and the superheated liquid flows from the inlet while the fins conduct the heat from the tube and transfer it to the air, blown by the fan through the radiator.

Cars operate on a wide range of temperatures, from below the freezing point to over 38 degrees Celsius (100°F). Water is an effective absorbent of heat but it freezes at too high a temperature for it to be work well in car engines. The fluid used in most cars nowadays is a blend of water and ethylene glycol, also known as anti freeze or engine coolant. This coolant also prevents corrosion and rust on the metal components.

How Long Does a Car Radiator Last?

Like most parts in an automobile, a radiator has a finite lifespan. Some car radiators last a few years while others last decades. Also, because of the location of a car's radiator, it is prone to damage even from minor accidents and fender benders. Break or leakage of a radiator renders it inoperable almost right away. Trying to fix cracks in the radiator may not prolong their life. Even solvents as strong as epoxy succumb to the high pressures and temperatures.

The Maintenance

Apart from the fact that it can completely ruin an engine, overheating reduces the mileage of a car. For maximum output and better performance it is necessary to maintain the radiator regularly. The following tips can help:

Correct coolant levels: Ensuring that the coolant is always topped up will help the radiator run smoothly. If air is trapped in the cooling system, bubbles will hinder the flow of the liquid. Also make sure you top up with the same type of coolant.

Check for leakage regularly: The coolant may leak at the hoses or the cap which would leave the radiator inoperable right away.

Check for clogging: Depending on the severity of clogging, the radiator might have to be cleaned properly or even replaced. Experts say, flushing may not be an effective solution to clear the clogs as it only removes the dirt and rust.

Using the right radiator fluid: The variety of coolants found at an auto supply store may confuse the average person. The right fluid for your car radiator depends on the climatic conditions in your region.

The best radiator fluid: Glycol based fluids mixed with as much water as the coolant works for most climatic conditions. In colder conditions however, a 70/30 ratio of coolant to water may be needed to prevent the engine from freezing off.

The Replacement

Several factors could account for a car radiators inoperability. One or more of the following parts of the cooling system may have to be replaced.

Thermostat: Severe heating problems can damage a good thermostat.

Water Pumps: The wrong pump or even a minor wobble in the right one could cause the engine to overheat.

Belts: A loose belt may prevent the pump from circulating the liquid fast enough and the fan from running fast for proper cooling.

Fan: The fan is an important part of the cooling system. Blowing cooler outside air through the radiator, if the fan does not operate appropriately temperatures may rise up to 50%.

When replacing a car radiator, several factors should be kept in mind. Due to the large variety of different types of radiators available in the market, users should have an idea of what to look for. Construction, tubing, airflow and cost are some of the things that affect the radiator. Aluminum radiators are the most preferred ones due to their resistance to corrosion. The toughness of aluminum also makes it a reliable material for use in car radiators.

Wider tubes ensure a larger surface area so that a larger volume of liquid can be carried, resulting in the liquid to be cooled faster. It is also preferable for the tubes to be not as thick as this can reduce the efficiency of the radiator. Electrical fans are more widely used due to their economical consumption of horsepower.

For replacing a car radiator, you require minimal tools but quite a bit of technical experience and there are several problems a professional can diagnose by instinct that an average person cannot.

CAR BASICS - Brake and Its fluid!

All that runs has to stop. You might have good stability, handling, comfort and power but when it comes the time to stop, you would like to be able to do that as better as possible. So stay tuned on the brake system because the focus should be redoubled with the Preventive Maintenance. It is composed of cables and cylinders, but the most important parts that are working effectively in the braking of the car are disks, pads and drum. These components act directly on the wheel of the car and make the car stop when you step on the pedal.

In general, discs and pads are located on the front, and are responsible for about 70 to 80% of the brake efficiency of a car. This set has better performance, as it dissipates more heat and water, in addition to simpler maintenance.

The drum set or canvas is housed in the rear wheel. The periodic review is recommended for every 5,000 km. But before that, depending on the use of the vehicle, the system may present some problem. The problems are: vibration and change of course when stopping, hand brake course too long, the pedal height (low and high) and constant noises when you step on the brake. Another component to be evaluated is the brake fluid.

As it absorbs moisture from the environment, its lifetime is about one year or 10,000 km. With time, the fluid has already a significant portion of water, reducing its efficiency in powering the brakes. The lack of fluid can cause the complete loss of brakes. The wear of the brake system and lack of preventive maintenance are some serious security problems. The least that can happen is the car need more space to stop or even, in extreme cases, do not brake in time when it is needed more.

Brake Fluid

The Basics

With a basic hydraulic system, brakes are designed to use kinetic energy to slow down a moving vehicle. Whether driving at slow speeds or high, by pushing down on your brake pedals your car will use a fluid to transmit a force which your tires will then use to create a friction that will ultimately slow things down.

Understanding the full extent of the technology behind brakes requires a great deal of knowledge but what even the most novice driver will know is that without your brakes your car will not function but how many are aware of what they can do to ensure their brakes remain in prime working condition?

The condition of your brake pads is essential and as a driver, it is crucial to ensure that you maintain and monitor regularly. This is not only to ensure the smooth operation of your vehicle but to improve your safety on the roads. One of the crucial ways of doing this is by knowing just what your brake fluid is, what it does and how you as a driver can use it to determine just how well your brakes are operating.

Whether you are new to it all, not sure of whether you know enough or just curious; below are the brake fluid basics that every driver needs to know…

- Brake fluid has many responsibilities but its primary focus is the corrosion protection and lubrication of brake systems.
- Without the fluid, your hydraulic brake system will not be able to operate.
- Brake fluid is held in a fluid reservoir.
- Adding fluid is generally not part of a standard vehicle service.
- Low fluid levels or a sudden drop can be an indicator of several issues including low brake pads that may need to be changed.
- On a few occasions air may enter the brake line. Bleeding the brakes can improve this condition however if your reservoir is struggling to contain the liquid then you may need to get it checked by your mechanic. This may not work with newer vehicles so check your manual before doing anything.

- Brake fluid must never be substituted with any other fluid.
- Become familiar with the brake reservoir so that you can easily check your fluid. There will be a "full" mark which will give you an idea of what level your fluid is at and what it should be at.
- If your fluid level falls below the "add" line then you may need to have your discs checked.

As the car owner you should have no issue in topping up your fluid however if you have any other issues or any general concerns over your brakes then you should see a professional.

Getting from a to b is important but doing so safely should be on top of your list! By knowing as much as possible about your brakes, how they function, what role the right fluid plays and any issues that can occur you have a great chance of ensuring your car remains in the best condition and you as the driver are as safe as possible.

How To Change Brake Fluid?

The hydraulic braking system of most cars and motorcycles use a particular type of hydraulic fluid which is called **_braking fluid_**. This fluid is very important in the maintenance of the vehicle like protecting the braking system from getting corroded and also keeping it lubricated. Thus, this fluid itself needs a bit of maintenance in exchange of all these. Replacing this brake fluid once a year, minimum is necessary to save some of the costly expenses one might have to bear if it is not replaced on time.

It is essential to keep the manual of your car or bike handy while changing its brake fluid. Motorcycle owners have to find the valve at the brake caliper and then take it out. Now, a small tube made of plastic, or the bleeding line has to be attached to the stopper. Check whether the other end of the line is inside the bucket, so that the used braking fluid can accumulate there. Now, the fluid reservoir is to be unscrewed with the required tools. A small cup will be seen either at the front or at the rear side of the bike depending upon which brake is bleeding. No moisture is to be let in during this process. Now, seep the used oil by pressing the required brake handle into the bucket. Add more fluid to the reservoir whenever the fluid level reaches the bottom, until the whole brown colored oil has seeped out, been replaced by golden colored new oil.

The procedure for replacing car brake fluid is similar to that of motorcycles. Another person has to sit inside it and keep the brake pedal pressed while, outside, you bleed the brake fluid. Once it is made sure that all the vents are secured, the brake pedal inside can be released slowly. The fluid oil may need to be refilled if the required fluid level is not reached inside the master cylinder.

CAR BASICS - Power Steering Fluid!

Inspection

Your car's power steering system is little more than a hydraulic system that applies pressure to one side or the other of the power steering rack or box.

It is driven off your engine's accessory belt and is typically a pretty reliable system as far as car's go. The only maintenance you generally need to do to it is to check to make sure the fluid is filled properly.

Below are the procedures;

Locate the power steering pump and reservoir. In most cars, the reservoir is built right into the pump and sits on top of it. In rare exceptions, the reservoir is separate, but it is never far from the pump. Just follow your engine's accessory belt around until you find the part that has a fluid container on it – that is your power steering pump.

Unscrew the cap and examine the dipstick or fill line to check the fluid level.

If low, add more fluid for the specific type of vehicle that you have. German manufacturer's such as VW typically like to use synthetic mineral oil, while most other manufacturer's use a more common hydraulic fluid. You can find this information in your car's owners manual or your local service department in the dealership.

How To Replace Power Steering Fluid

System components inevitably begin to degrade, over time. Metal may rust, and o-rings may deteriorate, releasing contaminants into fluid. These tiny contaminants may begin to clog the power steering pump, causing components to corrode. A visual inspection of the fluid's color will help car owners to determine whether changing fluid is necessary. Clean fluids are red, orange, pink, or somewhat transparent, while dirty fluids are black, brown, or impossible to see through. Car owners should check their owner's manual, for the appropriate maintenance interval.

Before flushing a system, the integrity of all parts should be checked. Checking for hose leaks, either from faulty hoses or faulty clamps, will protect the integrity of the newly replaced fluids. Also, if the car utilizes a cooling coil or return line, which is exposed to road salts or debris, the line should be replaced, if significant damage or corrosion is noted.

The simple flush method does not require the use of a jack. Using a syringe or gear oil pump, as much liquid as possible should be removed from the reservoir. Some mechanics recommend using a turkey baster, but the baster may not seal well, causing liquid to drip all over the engine. Once old liquid is cleared out, the reservoir should be filled to a level between the maximum and minimum line. Then, car owners should start the car, and let it run for approximately two minutes. This process should be repeated about six times, until about two quarts of liquid have been cycled through the system. While this system is easy, old fluid is never entirely removed from the system, increasing the number of times that fluids may become contaminated.

Flushing old liquid completely is more complicated, but also more effective. Car owners should start by removing their power steering filter, and then detaching the overflow tank from the radiator. Then, car owners should remove the reservoir hose, and replace with another hose, which should lead to an empty container. Next, after jacking up the front of the car, new lubricant should be poured into the system, which will force the old lubricant out, through the hose, and into the container. When the liquid from the hose runs clear, all of the old lubricant has been removed from the system.

The replacement hose should be disconnected from the reservoir. Then, car owners should replace the original hose, and pour clean lubricant into the reservoir. Racking the steering wheel a couple of times will squeeze out any air which may have become trapped in the system. After air has been removed, the overflow tank and filter should be put back into place.

A generic lubricant should never be used in a power steering system. A generic lubricant may be less expensive, but may also damage the rubber inside the car's steering mechanism. Some power steering fluids on the market are substandard, and may not contain enough anti-foaming agents, friction modifiers, or viscosity improvers. Using substandard fluids may overheat the system, causing breakdown of bushing and failure to hold pressure. Consulting the car's owners manual, or an original equipment data system, will let owners know which lubricant is right for a particular car's system.

The steering mechanism is one of the car's most important components. Therefore, if car owners doubt their mechanical expertise, they should consider hiring a professional to flush the old lubricant from their system. While replacing power steering fluid is not one of the most frequent maintenance procedures, replacing dirty lubricant will extend the life of the steering system, and keep the car driving normally for years to come.

CAR BASICS - Windshield Power Fluid!(*All you need to know*)

Have you been planning a road trip with the family? If so, or even if not, it always makes sense to assess the condition of your vehicle. Whether you are driving across the country or across town, a few minutes spent now to check the condition of your car or truck can save lots of headaches, time, and money later. A mechanical problem in the middle of nowhere or even in the middle of somewhere! -- is no fun and can even be dangerous.

If you live where it snows, consider what ice, snow, salt, and loose gravel can do to your car. The radiator can corrode without you even knowing it, minor scratches and nicks can turn rusty, and the windshield is subjected to the extremes of interior heat and exterior cold. These factors, and others, affect your vehicle when the weather changes from winter cold to summer hot.

A windshield that has a rock chip or ding at the end of winter needs to be repaired before thermal shock causes the minor ding to become a major crack. Thermal shock can also happen when cold air from air conditioning blows directly on a windshield that is very hot from summer sun exposure.

It is ideal to repair a windshield, rather than replace it, whenever possible. Windshield glass repair saves the windshield, and preserves the factory's safety seal of windshield to auto body. Because passenger side air bags deploy off of the windshield, preserving the factory installation is an important safety consideration. It also helps to avoid air and water leaks.

Another way to take care of your vehicle and prevent performance problems is to check fluids. Checking fluids is the cheapest and most important proactive maintenance you can do for your car or truck. Oil should be changed frequently every 3,000 to 5,000 miles and if you haul heavy loads or drive in "stop and go" traffic.

Also, flush your radiator and change your engine coolant every two years. The summer season can be very hard on a vehicle's cooling system so inspect your radiator for signs of leaking or corrosion. If you are not sure, have your mechanic check the radiator core to make sure it is not plugged or at risk of imminent failure. Check and fill other fluids necessary for your vehicle's performance, to recommended levels. These may include power steering, transmission, brake, radiator, and battery. Also, top off windshield wiper fluid. Do not wait until you need it, to do that!

CAR BASICS –*Ignition!*

One of the more significant parts to ensuring the smooth operation of the vehicle includes the ignition system. The ignition is able to trigger the engine cycle that is designed to drive the vehicle forward. If you wish to avoid starting problems with the car you want to make certain that the ignition is kept in full working condition.

If you fail to keep on top of maintenance of the vehicle, especially with components like the ignition system, you might soon learn that you are left stranded by the roadside due to complete failure. Basically there is a variety of problems that could cause the ignition problems, but since several components are related to the ignition system, the issues are often rather complex to diagnose.

Starting and ignition systems are vital for the car and they both require electricity to operate.

Troubleshooting electrical problems

Here are some guidelines for troubleshooting:

One should make sure that the automatic transmission is in the correct gear and the clutch on a manual transmission is completely depressed. If the car does not make a noise when it is started, check the battery terminal connections and then look for other loose connections in the ignition systems.

Replacing the Car's Starter

For replacing the starter motor or starter solenoid, follow these steps:

- First, remove the negative cable from the battery.
- Locate the starter, it is a nearly three inches round motor located at one side or the other of the flywheel. The solenoid is probably mounted on the side of the starter.
- Disconnect the battery cable and other wires attached to the starter or solenoid.

- Remove the solenoid and starter by removing two bolts that mount the starter. Be careful not to drop the starter. In case the solenoid is not attached to the starter then remove it from the firewall.
- Replace or repair the starter and solenoid. It is better to buy a replacement from an auto dealer. It should be tested before buying.
- Re-install the solenoid and starter. All the connections should be tightened and checked before using the starter.

Diagnosing and Treating Ignition Ills

The job of an automotive ignition system is simple, it supplies a spark to the engine at the time it is most needed. Fortunately today's technology has stabilized and the newest cars at least have some logic. In addition, more modularized systems are installed in cars. Not even mechanics repair ignition systems, they replace bad components. By using a simple volt-ohmmeter, many problems can be tracked down and many ignition system problems can be solved.

Repairing Ignition switch

The ignition switch of today has become more complex, earlier it was simple, three-position switch with Off, On and Start. The ignition switch is now linked to sensors, interlocks, anti-theft devices and the bank. The failure of an ignition switch typically is fortunately traced to a loose wire. This can be fixed if you can find it.

For repairing an ignition switch and wiring, follow this simple guide:

1. Locate the electrical schematic of your car's ignition system. The schematic tells what is in the ignition wiring system such as sensors and interlocks, besides the switch. Usually it is printed in the car manual or an aftermarket service manual.

2. Visually inspect the ignition switch and wiring for any loose wires, burns or any other damage. It can be reconnected and replaced as needed.

3. An ohmmeter can be used to test the continuity of wiring and ignition switch. The defective parts have to be replaced as needed.

Replacing Sensors and Control Modules and Distributors

A car's ignition system has many electrical components like sensors, control module and distributor.

For testing and replacing electronic ignition components, follow these steps:

1. Locate the car's electronic ignition and distributor. It is also called ignition control module or ICM. It controls the ignition system and is mounted either within the distributor or nearby.

2. Move the ICM cover as needed, inspect it for any problems such as loose wires, cracked cap or rotor. It has to be cleaned and replaced as necessary.

3. Use an ohmmeter for testing continuity of each component. Locate the sensors and test it along with the ignition control module.

4. Remove and replace the distributor if necessary. Note the rotor's exact position so it will help in reinstalling the new distributor with the rotor in the same position.

5. Document the required steps very carefully for easy reference.

CAR BASICS - Steering And Suspension!

Most car owners and drivers do not know a lot about their car's steering and suspension system. Suspension, when discussing cars, refers to the use of front and rear springs to suspend a vehicle's "sprung" weight. The springs used on today's cars and trucks are constructed in a variety of types, shapes, sizes, rates, and capacities. Types include leaf springs, coil springs, air springs, and torsion bars. These are used in sets of four for each vehicle, or they may be paired off in various combinations and are attached by several different mounting techniques.

The suspension system also includes shocks and or struts, and sway bars. Back in the earliest days of automobile development, when most of the car's weight (including the engine) was on the rear axle, steering was a simple matter of turning a tiller that pivoted the entire front axle. When the engine was moved to the front of the car, complex steering systems had to evolve. The modern automobile has come a long way since the days when "being self-propelled" been enough to satisfy the car owner. Improvements in suspension and steering, increased strength and durability of components, and advances in tire design and construction have made large contributions to riding comfort and to safe driving.

The suspension system has two basic functions, to keep the car's wheels in firm contact with the road and to provide a comfortable ride for the passengers. A lot of the system's work is done by the springs. Under normal conditions, the springs support the body of the car evenly by compressing and rebounding with every up-and-down movement. This up-and-down movement, however, causes bouncing and swaying after each bump and is very uncomfortable to the passenger. These undesirable effects are reduced by the shock absorbers.

CAR BASICS –*Electrical!*

Identifying Car Electrical Problems

Even though gasoline is the fuel that most vehicles use, each vehicle has an extensive electrical system that not only starts the car, but powers and runs all of the electrical systems. If any component of this complex electrical systems fails of gets faulty, your vehicle could get stopped in its tracks. You can not go very far with a faulty alternator or a dead battery. In any vehicle, there are at least four different components that are needed just to start the car. If any of these systems gets faulty or fails, you will not be going anywhere until you fix the problem. The purpose of this book is to help you identify and diagnose the particular electrical problem you might be having and a few tips on what you can do to solve the problem.

When you put the key in the ignition and turn the key, the ignition switch sends a signal to the solenoid and the solenoid closes the circuit that sits between the battery and the starter, which causes a large amount of voltage to flow out from the battery and to the starter to turn your engine over and start the car. After that happens, the alternator takes over the powering of most of the electrical systems and also begins to quickly recharge the battery for the next time you will need to turn on the car. Your coil and spark plugs give the spark for combustion to light the fuel air mixture that is in the combustion chamber to run the car as far as it needs to go. If any one of the systems gets faulty or breaks down, you might have some large problems to deal with.

So here are a few things to note for finding out which electrical problem you might be having.

If your car is completely unresponsive when you turn the key, but your headlights still work, then you most likely have either a bad starter or solenoid. The battery still has power because your headlights still function, but the power can not get to the engine to turn it over because of a bad starter or solenoid. If the car is unresponsive when you turn the key and the headlights do not come on, then you might have a loose connection on the cables that attach to your battery, or you may have enough corrosion that power can not flow from your battery.

If your car responds when you turn the key but you do not have enough power to turn on the car, then you most likely have a bad battery. This is not necessarily indicative of a completely bad battery, because at times, if you let a car sit idle for a period of longer than a week, the battery will start to lose voltage. This does not mean that the battery is bad and it should run just fine once you jump start your car and recharge the battery. If you have not let your car sit idle, and have noticed that it is becoming increasingly difficult to start your car, then you will probably need a new battery. Batteries wear out every few years and will not hold a good charge anymore. If you have not bought a new battery in a few years and you are having trouble starting your vehicle, you most likely will just need to buy a new battery.

CAR BASICS- Avoiding Car Accident!

- ✓ ***You do not need to be an F1 driver to drive well***

Every day there are hundreds of accidents on our roads that could so easily have been avoided. Insurance premiums go up for every bump or scrape we inflict on each other and our roads are only going to get more and more congested so what can you do to avoid having an accident that might just make your car not worth keeping anymore?

Here are some suggestions;

- ***Due care and attention***

Of course the biggest liability in your car is you. The majority of accidents are caused by the drivers who are not paying due care and attention to what they are doing on the roads. We have all done it: found ourselves trying to tie our shoe lace while eating a sandwich while on the phone and singing and trying to read the headline on a newsstand as you drive down the high street.This is applicable virtually to all potential drivers.

When you drive, you need to remind yourself that you are driving a lethal weapon. Just one slip on your part could mean someone or even you lose your life so being alert and ready for anything will help you to stay safe.

- ***Intoxication***

Quite rightly, driving while under the influence of anything is illegal. Even driving while sleepy is a bad idea as it can be just as dangerous as driving while drunk. Being intoxicated will mean that you do not have full control of yourself, let alone a powerful metal box that can go 100mph, so try and avoid it as best you can.

- *Maintenance*

A fundamental thing you can do to avoid having an accident today is to make sure your car is maintained properly. You might be completely sober and at the height of your awareness, but if your brakes do not work there is nothing much you can do to stop yourself driving into the back of someone. While your MOT should take care of all your cars basic elements for safety such as brakes, tyres, lights and steering, you can always get some performance car parts like special brakes which will give you that added level of mechanical safety.

- *Confidence*

Once you have all of the above sorted, the only thing you can do is to improve your confidence. Slow and tentative drivers can be as much of a hazard on the road as people that drive too fast. More lessons and more driving in tough situations should help you to do this and will hopefully mean you always get home safely.

CAR BASICS - Emergency Road Services!

What You Need To Know

Have you ever wondered what you would do if you were caught in some technical bind while driving your car? Before you step out of the house, keep the contact of a reliable emergency road services provider who can rush help to your side if and when needed.

There have been times when we have had to pull over due to some car trouble or failure. Some of the problems include smoke coming out of the hood or shortage of gas or even light problems. Why, some of us have also been victims of minor accidents. When in such situations, you should be aware of what the need is so that when you call in for help, you can give precise instructions to the service provider. This will empower them to dispatch the required assistance to you without wasting time.

When you have a burst tire, failed engine, dead battery, in such situations one should call for towing truck. They make it a point to instantly come to the spot so that your car can be pulled out of the way of continuing traffic. Depending on the nature of repairs, they will leave if the vehicle is fixed there itself or will take the car and you to the nearest service garage. Such a service is called vehicle recovery.

If your vehicle is stuck middle of the road due to tire burst, engine failure, battery exhaustion or any other reason, one should call the towing services. Towing services immediately respond and solve traffic snarls by initially pulling your car to the side of the road. If the car can be repaired right there on the spot, they will move out. However, if the car requires to be hauled to the garage, they can transport the vehicle there with ease. This is known as vehicle recovery.

When there is failure in the mechanism of the car, ensure that they send over a certified car mechanic to the scene so that the problem can be solved effectively. Most of the technicians have general spare parts with them once they know what the problem is and what is required to fix it. For example, if your battery has died, they will have the wires and mechanisms to revive it temporarily.

If you are making a long trip, it is inevitable that you will run out of gas. When such things happen, more often than not, it is tough to locate a gas station with ease. You would also not want to abandon your car in the middle of the road. Let the service know what you need, try to give as precise location directions as possible. This will help them to get to you fast. They would also be able to know if there is any station around and if not they could bring the fuel themselves to solve the situation. You too could suggest this to them.

Long journeys often render a driver out of fuel and sometimes with no gas station in sight for miles. When one reports this to the car emergency service, they will calculate the distance of your location to the nearest station. If it is possible to tow the car in, they will send a tow vehicle but if it is too out of the way, they have the means to send in mobile refueling services. Other minor car emergency services include water delivery. Water for burnt out carburetors as well as for your drinking purposes can be availed of. Another service that most emergency service providers give has nothing to do with the failure of the car. They also provide relief for mistaken lockouts and alternative modes of transport if the car cannot be opened through duplicate keys.

Medical aid, on the spot injury and ambulance facilities can also be availed of if required. Most emergency road services offer credit facility and accept credit card payments. It is very important to opt for good road emergency services that offer quality and human care.

CAR BASICS - Road Safety Tips!

- ***Avoid caffeine and sugary snacks***

Although you may get a burst of energy from a sugar high, it is short lived and you will feel more tired afterward. Choose a snack such as dried fruit, nuts, or a muesli bar which have a low GI, and drink plenty of water.

- ***Have backup shoes or drive bare foot***

Your foot is more likely to accidentally slip off the pedal if you are wearing thongs or high heels. Despite common belief, it is legal to drive barefoot - if you are wearing slippery shoes take them off before you drive, or keep a backup pair of shoes in your vehicle that have good grip.

- ***Avoid standing on diesel spots***

Even if you drive a petrol vehicle, diesel pumps are often side by side which means there may be diesel spillage nearby. Diesel is oilier than unleaded and could leave the soles of your shoes greasy, causing your foot to slip off the pedal.

- ❖ ***Do not swerve to avoid hitting an animal***

As awful as it is to think about striking and potentially killing an animal with your vehicle, police advise it is safer to hit an animal than swerving to avoid it, which could easily lead to a serious crash. Consider protecting yourself and your vehicle with a bull or nudge bar.

- ❖ ***Do not be a hangover drink driver***

Surprisingly, 25% of drink drivers are caught between 5am and midday. This could happen quite easily if you have had a heavy night of drinking, went to sleep late and got up early the next day to pick up your vehicle or drive to work. Ensure you give yourself plenty of time after your last drink for the alcohol to pass through your system.

❖ *Be careful when changing a spare tyre*

Unfortunately each year there are accidents involving drivers who are hit whilst changing a tyre. Ensure you drive your vehicle slowly off the road, away from traffic, or if you are on a highway, try to reverse your vehicle behind a safety barrier.

❖ *Keep your headlights on, day or night*

Drive with your headlights on at all times to increase your chances of being seen, especially in areas with high pedestrian traffic or long, shady stretches.

❖ *Missed your turn off? Go around*

Do not take a chance on causing an accident by trying to cut across lanes to make the turn off. Take the next exit or turn and enjoy the scenery.

❖ ***Do not overdo it!***

Do not overload your vehicle's roof rack - most cars can only handle between 60 to 100kg up there. Any heavier and it could damage the roof rails and start to affect how your car handles corners due to the raised centre of gravity.

❖ ***Check your tyre tread***

If your tyres have less than 1.4mm of tread depth you could find yourself in trouble - with the law or on the road. Many people measure tyre tread in the middle of the tyre, but if the edges are worn down further than 1.4mm, the vehicle is considered to be unroadworthy and dangerous. Wheel alignments and maintaining the correct tyre pressure will provide you with better vehicle handling and will increase the lifespan of your tyres. Consider purchasing an aftermarket tyre pressure monitoring system (TPMS).

- ❖ ***Check your trailer axles***

Likely your trailer or caravan has been sitting in the weather for close to a year. Get your axles and bearings checked and regreased or replaced - there is a good chance the axles may be rusty or the bearings seized.

- ❖ ***Add emergency details on your phone***

If you have an Apple iPhone, you can record emergency and health related details using the built-in Health app. This means if you are involved in an accident, emergency workers are able to access information about your allergies, blood type, and emergency contacts, even if your phone is locked. Android users can download this App.

CAR BASICS - Emergency Items To Keep!

We all hear about car accidents and breakdowns and think that it would never happen to us. However, what if it did and you were not prepared? It is always better to be safe than sorry, so making sure you have an emergency kit in your car is advisable as you never can tell what might happen and when.

Emergency kits have helped millions of people over the years and at the same time millions have wished they had one. If you break down in a remote place where there is nobody to come and assist you quickly or where it would take the emergency services a long time to reach you, then you will benefit from carrying an emergency kit and being prepared.

The first thing to have on board is a first-aid box; these can be brought from most pharmacies or large supermarkets. Some are more comprehensive than others but you should make sure that the kit has at least painkillers, bandages and plasters, wet wipes and any medication you need to take regularly.

In the event you are unable to drive your car you should have the necessary tools in the boot to help you fix it. Typically the tools you should have with you are a tyre iron, car jack and possibly screwdrivers for electrical faults, although this is best left for roadside recovery or breakdown services to deal with. Having jump leads on board is essential should your battery go flat.

All of these items are useful should you be unfortunate enough to be involved in an accident. Duct tape is also brilliant to have as it serves many purposes such as makeshift repairs of bodywork or bumpers. If you can get hold of a high-visibility jacket then do so as they are very useful if your car breaks down at night. In poor visibility or at night it is important that you make your presence known to other motorists to avoid any accidents.

Other items you should consider having are blankets in case you break down in colder weather, drinking water to keep you hydrated (or to fill washers should you run dry). A flashlight or torch is handy so you can see what you are doing if you break down at night.

The best thing to do with emergency kits is to leave them underneath the passenger seat, in the glove compartment, in the boot or with the spare wheel so that it is out of the way but easily accessible should you need it. Try and stay calm and remain in your car if possible and wait for help to arrive.

The majority of us spend a significant amount of time in our cars or vehicles, therefore it is wise to keep certain items safely stored away in case of an unexpected emergency.

If you need to venture outside then make sure you are careful and be aware of your surroundings at all times, especially at the road side or on a hard shoulder. Finally, it is advisable to carry a mobile phone with you when on any journey and to make sure that your phone has plenty of charge in case of an emergency.

Below are some of the highlight of emergency kits discussed above and others not discussed.

- **_Flashlight with spare batteries_**

If you break down or get a flat tire at night, a flashlight is an indispensable item. Without one, checking engine components or changing a flat can be a very difficult and frustrating undertaking. In addition, be sure to always keep a number of spare batteries for the flashlight - you do not want it dying when it is needed most.

- **_First Aid Kit_**

Injuries, whether major or minor, can never be foreseen, and you never can tell what might happen while you are out on the road. Be prepared for anything from scraped knuckles to small cuts or a bee sting. Basic emergency first aid kits are available at most department stores and pharmacies, and should contain most everything you may need to treat an injury.

- ***Bottled Water***

In your travels, you may find yourself broke down or stranded in a less than desirable location. Having water available ensures you can keep yourself and your family members hydrated.

- ***Glass Scraper***

If your area normally sees snow or ice during the winter months, an ice scraper for your windows and windshield is essential. Even in these fall months leading up to winter, it is best to be prepared now rather than unprepared when you actually need it. Be sure to keep a scraper handy, you never know when you might awake to find your vehicle covered in ice or frost.

- ***Jumper Cables***

At one point or another, most of us have had the misfortune of a dead battery, and find that we did not have battery jumper cables in our vehicle. Hearing that empty click when you turn the key can be a real downer, but if you happen to have a set of cables you can quickly get a jump off another considerate motorist. Having these cables in your vehicle also enables you to help someone else out in a similar situation.

- ***Hand Warmers***

These very small yet useful items can be easily fit into your first aid kit or glove compartment. For those of you who have changed a tire in extremely cold weather, you will understand why we recommend these as a must-have. If you have not- take our advice and pick up a pair! Working around your vehicle in the biting cold without hand warmers can be an extremely uncomfortable experience.

- ***Duct Tape***

As the saying goes, "Duct tape can fix anything". While that may not be entirely true, having a role of duct tape handy is always a good idea, providing a quick, temporary fix to an unforeseen problem.

- ***Hazard Triangles or Flares***

These items are for the safety of both yourself and oncoming motorists. If you break down on the side of the road, placing hazard triangles or flares behind your vehicle will alert other drivers that there is a vehicle in distress up ahead. Without these items, other drivers may not see you in inclement weather, resulting in a near-miss or even a collision. Place these items at 50, 75, and 100 feet to the rear of your vehicle, and on the side of your vehicle facing traffic.

- ***Tow Rope***

Ensure the rope you buy is capable of towing your vehicle's gross weight (see the owner's manual to determine this figure). Keep in mind this rope is not for extended trips and regular towing, but only to get out of harm's way.

- ***Small Shovel***

Many of us at some point have found ourselves stuck in some type of mud or snow. Having a small shovel handy will allow you to dig out around your tires if needed. Also, you can use the shovel to pack material in front of your tires for better traction when trying to free your vehicle.

CAR BASICS -Preparing For Emergency!

With the amount of driving we do today it is only but wise to ensure safety on the road. It is best to be prepared and prepared for a road emergency at all times. Having the right tools and supplies can make all the difference in getting through the ordeal safely and with the minimum of hassle.

1.The best way to prepare for car emergencies is to first reduce the likelihood of emergencies occurring in the first place. You can help the cause greatly by making sure all your car basics have been attended to like checking the oil levels, radiator fluid, coolant, washer fluid, making sure the tire pressure is at the proper level, and making sure there is ample amount of tread on the tires.

2.Ensure that you have a spare tire and that it is properly inflated. Along with a working spare tire be sure that you also have a working jack, tire iron, a short galvanized pipe to use as leverage for the time iron. A can of "instant sealant" can be a good addition to your emergency kit.

3.A strong flashlight and road flares can help to keep other drivers alert of your situation.

4.It is also a good idea to keep some provisions such as freeze-dried food and water as well as cold-weather gear such as blankets and parkas.

5.A first aid kit consisting of gauze, antiseptic spray, bandages, tape, scissors.

6.A car tool kit with the basic tools such as flat head and Phillips screwdrivers, pliers, adjustable wrench, wire cutter, spare batteries, work gloves, spare fuses, vice grips, duct tape, box cutter or utility knife, electrical tape and scissors. Also be sure to have jumper cables.

Keep your emergency kit items in a plastic waterproof box or heavy duty nylon bag. This will help to keep all the items organized and easy to find.

CAR BASICS -Benefits Of Emergency Kits!

These days we are so dependent on electricity. It is hard to notice during the day to day activities because we are so used to it. However, if you think about a time when the power goes out, are you prepared? You will not be able to see if it is dark outside or the room does not have any windows. Also, if the power goes out because of an emergency or natural disaster like an earthquake, flood, tornado, or the like, it is quite possible you will need a way to heal injuries.

There really are so many benefits to having and keeping emergency kits or tools for your safety and for the safety of your entire family. The problem lies in the fact that many people are just not prepared for a natural disaster to happen or for something as simple as the electricity to go out. In cases like these, having emergency supplies can help to get over this particular time a lot easier than if you were not first prepared for it. Having at-home emergency kits or equipment supplies can definitely come in handy while you are home and something happens. Instead of fumbling around for a drawer that contains these items, you can just reach for the kit that has everything handy.

Some good ideas for items to include in your kit can be candles and flashlights. These are helpful because without electricity, we can not see. Candles may seem old fashioned, but if you run out of battery power, they will start to seem invaluable. A first aid kit is another good idea because if you get injured in a disaster, you will be able to care for them.

Alternatively, there are emergency equipment kits available that are great for car use. If you happen to be having car trouble while on the road, it can be very frightening to not be prepared for it. The kit in question may include items that allow you to work on your car at night without worrying that you will be hit by a passing vehicle. The items included in this particular kit may be flares and cones that you can use on the side of the road. If you are constantly driving at night, one of these kits can become your best friend in times of need and it can even save your life, considering how dangerous it can be to work on a car's mechanics when the sun is not up.

CAR BASICS - Advantages Of Emergency Road Services!

An aware and smart driver knows how important it is to know about emergency road services. They are like Good Samaritan for a vehicle owner in need. Roadside breakdowns may give nervous down experiences where only emergency services come to help and protect a driver, his or her family members or any other companions. Towing services can help drivers or customers in case of immobilized automobiles with quick, efficient and affordable assistance.

No matter how hard anyone tries, he or she may experience anything unexpected when they are out, driving for long or short distances. Even a good vehicle may go wrong either with a blown out tire, losing a key or locking a key itself in the vehicle or running out of fuel. This is just a partial list of common vehicle problems and it can grow further with many more problems which causes a vehicle to get disabled. However, there is nothing to worry, as emergency road services across the globe are ready to help anywhere, anytime.

It is important to register with an emergency road service as there are unpredictable circumstances can come along the way. No one would like to be trapped out in open space, without knowing how and where to go. To avoid any such situations it is necessary to find out a good roadside assistance and getting ready for any misfortunes.

Professional roadside assistance is most reliable and very fast to provide needed service. It is just a phone call away when it is required and generally takes 45 to 60 minutes to reach the site. Whether a vehicle is stuck in or out of the city of the particular place, these road services answer it quickly if given a call as they record client's information and begin dispatching as soon as possible with help of their wide network. This service is available for everyday.

Emergency service representatives offer prompt services should any vehicle require towing service.

When an automobile goes immobilized or breaks down for any reasons, these services make it possible to get that in to working condition within hours. Towing will be damage free, honest and professional. They know the client's time is valuable and they try to satisfy him or her with professionally trained technicians.

Emergency road services and towing services are offered at an affordable price, and most of them charge small coordination fee per call. Convenient rate plans offer different choices to a customer and generally there is no sign up fee. Most of the road services take care of all types of vehicles, however some of them specialized in specific vehicles such as RVs. So when in need of road side assistance for such vehicles it is good to know where to find it.

A driver in need knows whom to call for emergency road services when he has registered with one of many good towing companies, so he does need to search out for help or to just wait on a road side. Rather he or she can drive with peace of mind that whenever a need arise a service will be right there for him or her.

CAR BASICS – Repair!

Knowledge about something is always an advantage.There will always be an event in our life that requires application of this knowledge. We must not really be an expert, but we should at least have enough idea of a concept, so that we will be able to utilize it. Cars are one of our life savers. It allows us to go home early and it frees us, from the hassle of commuting using public transportation system. However, cars might give us a big problem, especially if it suddenly stopped in the middle of the road and we do not have a single idea how to fix it or check what went wrong. This type of situation makes us realize how important it is to know the basics of car repair.

A recent study conducted revealed that many motorists have deficiency in car repair skills, even just the basic ones. This circumstance leads to more car breakdown that compromises safety of the commuter. Lack of skills in basic car repair also means greater expenses, since most of the simple activities like tire changing or oil top up are saved up for garages. Cars which are not properly maintained will also incur larger damage.

One does not have to enrol in a mechanic class to know the basics; although it can also be an option, depending on a person's choice. Maybe you can spend a day in a garage and you will realize that simply watching other people doing these simple tasks for a couple of occasions, will be enough to give you a gist of how they actually do it. Observation alone might not just be sufficient, so prepare to ask question, especially if things are unclear to you. For sure, they will be glad to answer your queries and share you their knowledge. Skills with tire or oil changing and basic tune ups will not be acquired just by observation. You have to make use of the concept you learned by applying it in actual, with the guidance of the expert at first. If you are already confident to do the stuff, then, it is already safe to do it alone.

One must have excellent diagnostic abilities in order to properly address the problem. Knowing the problem in the first place will lead you to the correct solution. Just checking the entire car starting with the hood will surely provide you with a good idea.

Furthermore, you cannot perform any task if you do not have the needed equipments. Make sure to have extra tire, car jack, towing cable, self illuminating signages and other car repair essentials in order to get the job done. You may also want to have a first aid kit in case any minor accident happens.

It is important for every driver to at least know how to do basic car repair. Every driver should have enough knowledge when it comes to repairing his own vehicle if there are minor problems because this will become useful during unexpected situations. It is not only beneficial because it will save you some money but also because you will be able to maintain your car in good condition.

The most important thing to do is to be aware of your car's performance and know which part of the vehicle really needs attention or repair. Noise is the fastest and the easiest way to find out where the trouble in your car is. Just pay attention to your car's performance while you are driving. If you hear an unfamiliar sound, it is most likely an indication that you have to check your car. When it comes to the electrical aspect, you will know if something is wrong when one of your lights gets busted or a particular function does not work. You also have to regularly check your battery because this is one of the most common problems for drivers and car owners. See if it is still functioning well or if it needs recharging or replacement.

Apart from those, there are some other basic car maintenance tips that every driver must know. Although you can always ask the expert to change the oil for your car, you should also understand how this goes. Oil changing is like giving your car engine the best food. It reduces the potential damages on your car's engine. The recommended time to change your car's oil is every three months or after you have reached 4000 miles. You also have to know how to check the fluids in your car. To avoid expensive repairs, you must regularly check your car's fluids for the power steering, brakes and transmission. You do not need to have a mechanic just to check it.

Every driver should know the basics when it comes to repairing simple car problems. This can save you a lot of time and money. It is just a matter of learning, practicing and you do not really need to have some formal education.

CAR BASICS- Auto Repair Tips!

A car is an important part of all our lives. It drives us to work, we use it for vacation, or we take it anywhere else we need to go. Just like with any other piece of equipment, however, cars will eventually break down and will need to be repaired.

Most people with a broken down automobile will immediately take their vehicle to a mechanic to get it fixed. Although this is a good solution, auto mechanics can charge expensive prices that can be avoided by fixing the car yourself.

Listed below are some helpful auto repair tips for fixing common car problems.

You should not be trying to take apart the engine if you do not have the experience or qualifications for fixing an automobile. But simple assessments and repairs can be done by almost anyone who has a basic understanding of how a car works. Some basic things that every driver should know is checking the oil, transmission, and brake fluid levels to make sure they are properly filled. Dirty oil may also need to get changed periodically.

Buying a car jack can be extremely helpful if you want to change tires yourself. You should always carry this device in the back of your vehicle in case your tire blows out in the middle of a road. Changing a tire can be very dangerous if not done properly, so try to take a short course on how to properly change a tire.

To keep your car's airflow working at optimum levels, it is important to regularly change the air filter. There are all types of junk that flies right into the engine compartment that can affect your car's performance if it is not cleaned off. A buildup of junk can even result in the "check engine" warning light appearing on the front dash of your car. If you have not changed the air filter for awhile, you may want to clean up under your hood at the same time.

When cleaning an engine, many people use methods that may be indirectly hurting their car. Many owners pop the hood of their truck and blast at their engine with a shiny cleaning solution. It is never good to blast your engine with a large amount of water, or any other type of liquid. Even a small amount of water inside the distributor cap will cause the engine not to start properly. If you need to clean your engine bay, do it carefully and follow the instructions provided by the manufacturer. If you feel like you do not know how to do it properly, just call an experienced mechanic.

CAR BASICS –GeneralMaintenance Tips!

Each one of us dreams of having our own car - may it be a simple one or a cool sports car. Owning a car sound great because it gives us convenience, comfort and joy but then owning one means additional responsibilities. Each car owner must know how to take good care of their unit from basic maintenance to simple troubleshooting. Car owners should treat their vehicles like their own child or partner. They should know how to maintain their car so it is always in good running condition and would last for a long time.

Achieving your goal of owning your own car is a great accomplishment. You are now free to easily reach anywhere that you want to go from the comfort of your own set of wheels. However before you start driving up and down the country you need to be aware that the task of owning a car does not end with just buying one, you need to know how to properly take care of your car as well.

Car maintenance, even in its most basic form is something that all car owners should be aware of. You need to be able to perform tasks such as checking the engine oil, the pressure of the tyres and the condition of your brakes. You also need to take your car in for regular service sessions and wash it on a regular basis in other to protect it from extreme weather conditions such as rain and snow. All of these represent the usual steps that all car lovers should take in order to maintain vehicles.

Being able to do aspects such as the above can save you money as you do not have to rely on the skills of a mechanic to do everything for you. All you will need is some basic tools to do the job and a bit of free time in which to do it in. So to help you achieve this see some of the maintenance facts and advice below:

Here are some basic car maintenance tips that car owners should follow and practice.

✓ **Check engine oil**

Checking the level of your engine oil is the most important thing you should consider before riding your car or traveling to far places. The car's engine consists of many moving parts which definitely need a good level of lubrication to protect them from untimely wear. There are two kinds of engine oil: a monograde which is the ordinary one and a multigrade which is a special formulated one with additives that protect your engine providing you more mileage before undergoing change oil.

✓ **Check engine level coolant**

Before leaving your place, it would be best to check your coolant level to prevent any high temperature problem that might lead to engine overheating. When your car starts, the process of engine combustion takes place and it reaches a thousand degrees, especially in the combustion chamber and that is when the engine coolant works. From the radiator, it passes through the coolant chamber inside the engine taking some of the heat off. As it passes through the different coolant or water chamber, it returns from the radiator completing the automotive cooling process. Some units use ordinary tap water but it would be best to use coolant to protect your engine. It is a mixture of ordinary tap water with a special formulated coolant that protect your radiator from what they usually call "scale" that may damages or clog your radiator.

- ***Make sure electrical system and tires are in great condition***

Make it a habit to check your electrical system and see to it that all important lights like head light, signal light, and park light are functioning well. This should be done to avoid car accidents and to avoid additional damage to your car. Checking your tires is also important before you go on a trip. Check for any flat tire and make sure to bring along a spare tire especially when going for a long trip.

✓ **Check car gauges**

Start your engine and check if the voltage gauge, fuel gauge, oil pressure gauge, air pressure and temperature gauge is in normal condition. For voltage gauge, after turning on your key you will notice that it reaches 24V but once you start your engine, it would increase from 24 to 28 Voltage, if not, there might be something wrong with your battery or your alternator. The fuel gauge gives you an idea whether you already need to refuel or if the fuel is still enough for the whole trip. It is hard to get stuck in the middle of the road just because you have not noticed that your fuel gauge is almost empty. As you start your engine your oil pressure automatically moves out but as soon as the engine heats, it would automatically would go down to its normal level and thus indicate that your car is in good condition. The temperature gauge shows the temperature of your engine. The normal temperature of an engine is one half of the gauge in flat lying area, but once you reach an inclined plane, your normal temperature reaches 3/4 of your gauge.

✓ *Check Battery And Brake System*

A car battery's life usually depends on how often the owner uses the vehicle and on the road he usually travels. Make sure to change your battery as soon as it is necessary to avoid further damage to the engine. You should also check your brake system and make sure that your brake fluid is still enough and that your brakes are really functioning so as to avoid accidents.

Brakes, in regards to safety these are one of the most important aspects of your car. It is compulsory that your brakes are kept in the best condition possible. However it is important to remember most brake problems develop gradually and do not rise to serious levels suddenly. Some of the main ways to spot potential problems within your brakes are through looking for aspects such as if there is a lot of pedal movement before the brakes begin to bite then they possibly need adjusting or if the brakes feel spongy or lack sharpness then it could be an indication of air in the system. Also if you pull on your handbrake and it takes more than a few clicks before it will hold the car on the hill then you need to get it checked out.

If you do feel there is a problem with your brakes it is important that you do not use the car until you get them fixed.

✓ **_Tyres_**

Your tyres have a legal minimum tread depth. If you drive on tyres that are in poor condition or that are incorrectly inflated it is dangerous and could potential cause you to have an accident, which is why it is important that you check the pressure of your tyres.

The right amount of pressure for your tyres should be listed in your manual. It is also common that your front and back tyres would be different amounts. When you are checking the pressure on your tyres you should ensure the tyres are cold otherwise you could get a falsely high reading. If the amount shown on the pressure gauge is below the amount mentioned in your owner's manual then you will have to inflate the tyres. If by mistake you put too much air in your tyres then you should depress the pin in the centre of the valve to let some out.

✓ Go to a car repair shop

Once you noticed that there is something wrong in your car and that even when you thoroughly checked your unit you can not seem to find the trouble, you should send your unit to a repair shop immediately. This should be done to avoid further damage to your car and to make sure that every part of your car is functioning well. Not only that, you should visit a repair shop once in a while for tire alignment, change oil and other basic maintenance services that are necessary for your unit.

These simple tips should be done to ensure that your car will last a long time and that road accidents can be avoided. Follow these tips and for sure you will enjoy the benefits you get for having a well-maintained car.

CAR BASICS – *Washing!*

Many people get confused by all the techno mumbo jumbo that car product manufacturers have been blasting out. Often they chase up to the newest and greatest compound rumored to not only bring out the best shine in a car but also one that reflects laser beams as well.

Like anything else in the world people forget about the basics and fall prey to exaggerated marketing efforts by car care product companies. To help you remember about what really needs to be done when you are to wash your car.

The below are some vital steps to wash car thoroughly;

- *Prewash*

Prewashing involves hosing down the body of the car with water. Sometimes you might need a lot of water to make sure you get the grime off very dirty vehicles. If your car is relatively clean then you will not need much time hosing down your vehicle. Rather you just wet it to form a base for the next step.

- ***Soap***

Soaping your vehicle usually is done with car soap and a mitt. The wrong way to soap a vehicle is by rubbing your mitt into the surface with force. Another mistake is also rubbing back and forth like polishing a jewelry piece. Car surfaces do not need to be rubbed and scrubbed like this. A gentle pass with your mitt over the surface of the car is enough to rid your car of dirt.

- ***Rinse***

Rinsing your vehicle can be quite easy. You just have to point your hose towards your car to rinse of the soap you used on step 2. Actually if you want it to be more effective you should hose down the roof then the windows. Afterwards you proceed to the body work and hood. Do not forget to rinse the wheel wells as well. If you have a pressure washer rinsing the tires will be much easier. Try and take care not to use too much pressure if you have an older car. This may cause chipping with heavily worn paint.

- **Dry**

This is arguably the most important step that people often purposely disregard doing. Drying a car makes it more resistant to dust. The dust and dirt that you spend a lot of time removing in the first three steps could build up faster without the proper drying procedure. What is more is that drying a car can prepare you for the optional next step which is waxing.

- **Waxing**

Waxing is a sealing process. By sealing we mean protecting the paint from outside forces that tend to ruin it. Sealing products include waxes and synthetic sealants which wrap your car in a protective skin that repels dust and grime. Not only that but this layer of protection acts as a buffer to the car surface. You can actually prevent scratching when using a good paint sealant.

Car Washing is very important because it removes contaminants that may otherwise eat up on to your paint.

CAR BASICS - What You Should Keep In Your Car When Going On Trips!

Going on a road trip can be fun and adventurous but you should be prepared for your travels. It is always better to be prepared for something to go wrong and not use it than actually run into a problem and need something you do not have while you are stuck on the side of the highway or on some long stretch of road where you have not seen a car in hours.

Below are some road trip basics to get you ready for your next journey.

✓ Ensure your vehicle is ready

First, before heading out on any trip, make sure your vehicle is ready. Make sure the tires are inflated to the right air pressure based on the recommended pressure given to you by the manufacture. Make sure that all of the fluids under the hood are topped off. It is also a good idea to get your oil changed before the trip if you are close to needing it done anyway. You will also want to double check and make sure your spare tire is full in case you were to need it on the trip. It is also good to start off your trip with a full tank of fuel.

Depending on how far you are going and the age of your vehicle, some people like to rent a car for their trip. This allows you to drive a newer car in most cases and you are not putting miles on yours which lowers the overall value. Most rental car companies offer specials based on the length of time you rent and nearly all give you unlimited mileage for your trip.

- ✓ *What you should keep in your car or take along for the trips!*

You want to make sure that you have everything you would need in case of a breakdown or some sort of emergency. Here is a basic checklist. You should prepare a seal-able container and keep these items in your trunk in case you would ever need them.

• ***Road Flares or Flashlight*** - it is always good to be prepared if you were to have problems at night. Road Flares and a flashlight will not only keep you safe but help you alert other drivers that you need assistance.

• ***Small Air Compressor*** - these actually come in handy for many things around the house but are useful on road trips as well if you have a flat tire. If you cannot change to the spare, you can in many cases get enough air in your tire to get to a service station where you can get help.

• ***Snacks-*** if using an airtight container, it is always good to pack some snacks in case you are on the side of the road waiting for help for a long period. Things like crackers are always good and should stay fresh for some time. I always pack a bottle of water for inside the car as well.

• ***Maps*** - many vehicles today are equipped with GPS and Navigation systems. If you are making a road trip in an older vehicle without this equipment, it is good to plan and print maps to your destination. I like to print maps of the area I am going as well so I know where to find restaurants and fun things to do.Remember road trips can be a lot of fun if you are prepared.

CAR BASICS – *The Buying!*

❖ *Do you really need a car?*

Before you buy a car ask yourself first, is a car really necessary? You should look at your current lifestyle if a car is really needed or not. For example, if you have a family then its almost a given that a car is needed. If you live alone and live close to a bus stop or a train station or your office is near enough, may be you do not really need a car. To others its more of a personal preference.

❖ *Are you financially ready?*

You need to look at your financial situation, because buying a car is a long term commitment and investment. From the day you purchased it to how long you are going to keep it, money will be involved. Think about it, your monthly payments, fuel cost, regular maintenance and unexpected breakdowns, warranties will only cover manufacturer's defect. If your planning to finance or lease the car, make sure that you have built up a good credit history for at least a year. Most banks or car dealers will not lend you any money if you do not have a good credit history. Some will but with very high interest. If you saved up some cash, put a down payment to reduce your monthly payments. You might find some good deals in the newspaper ads, but be careful, make sure you read the fine prints.

❖ **_Ready to shop around?_**

When you have finally decided that you need a car, then it is time to choose one. Make a list of what you want and need in a car. The internet is the perfect place to search. A lot of new and used car dealers and private sales have websites so that buyers can search for a car without too much hassle. If you see a car online that appeals to you, call the car dealer or the private owner and ask about the car. Take a look at the car personally, do not jump into anything just because the car looks nice outside, make the necessary inspections and research. Just make sure to check for any lien on the vehicle if your buying from a private party. An auto mall is also a good place to shop if there is one in your city, because all the different dealers are there, even auto insurance.

❖ *Time to buy the car*

When you finally found the right one, then this is it, this is time to sit down with the salesman or a private owner and do the necessary paper works, Remember to ask the questions and concerns you have about the car and consult with experienced buyers of new or used cars as to how to make a good choice and how to get about the sale. If you know a mechanic personally or someone who has knowledge about cars, bring him or her along.

❖ *Other sources of buying a car*

There are other ways of buying car, repossessed cars are very cost effective. Buying a repossessed car means that you can save a lot of money. If you do not mind owning a repossessed car. Government or public auctions are another way. There is a wide variety of vehicles, ranging from trucks, SUV's and even motorcycles. These vehicles are seized by the government or some financial institutions and belongs usually to individuals who can no longer payoff their debt.

CAR BASICS - Buying A Used Car!

Though you can almost always save money on a used car purchase, you need to be ready to search as a savvy consumer. Whether you are dealing with a private sale or a certified dealership, you should keep your eyes open for good deals and discipline yourself to shy away from bad ones. If you are considering the purchase of a used car, keep reading to learn the basics of a wise search.

> ### *Know The Car You Want*

The used car market is huge, so having at least a general idea of what you are looking for will really help you narrow your search and allow you to get a clear idea of pricing.

Once you have chosen a make and a model, do not just sniff around at dealerships, but watch your local classifieds. Also, put the word out to friends and family that you are looking. You never know when a good deal can arise through a private sale.

➢ Find Your Price Point

Once you have a few vehicle choices in mind, research their actual resale value. Using the Kelley Blue Book (available at kbb.com), you can look up a car's standard sale price using its make, model, year, mileage and condition.

While the Blue Book value is not hard and fast, it offers a basic guideline that will help you recognize a good deal and know when the price of another is just too high for what you had been buying.

➢ Before You Buy, Get The Papers

Whether you are buying privately or through a used car dealership, always ask to see the car's title search and even vehicle history report. You can obtain your own vehicle history report through a service like CARFAX that will tell you the car's sale history, past odometer readings, emission inspections, and major incidents.

If you are buying privately, ask the seller for copies of the vehicle's maintenance receipts. While not all owners will have full records, they should have some. Though such records certainly do not prove the vehicle is currently in good shape, they do provide a bit of additional evidence that the owner may have taken good care of it over the years.

➢ *Ask For An Inspection*

Unless you are buying a used car that comes with a manufacturer's warranty, always have a car inspected by a third-party, certified mechanic. Getting their independent opinion on the car can help you make a wise purchasing decision.

➢ *Always Test Drive*

Unless the car is not drivable or road-safe, it is critical that you take it for a test drive. You will get a feel for the car and a much better awareness of any potential problems.

So, if you are planning to buy a used car, remember to narrow your selection, have an idea of the Kelley Blue Book value, obtain the appropriate paperwork, secure an inspection and always take it for a test drive.

CAR BASICS – The Insurance!

When we think about car insurance, we usually think about the basic coverages that are required by law, such as liability coverage for both bodily injury and property damage. Collision insurance and comprehensive coverage are usually required if you have an auto loan with the bank. Some states require uninsured and under-insured motorist bodily injury and property damage coverage to fill in the gaps in case the person who causes the accident has less than adequate coverage. However there are quite a few other coverage types that you should know about and may want to include in your next auto insurance policy.

Medical payments insurance covers medical expenses for both you and any passengers that are in your vehicle who are injured in an accident. This insurance does not care who is at fault for the accident. This coverage may extend to you as a pedestrian who gets hit by a vehicle or even to you and your family if you are in someone else's car. This policy can be tailored to your needs, and is generally helpful if you and your family do not have a health insurance policy that covers these types of expenses. Personal Injury Protection or PIP is a similar type of policy.

Another very important type of coverage to consider is Work loss, because it helps you get back lost wages that are directly related to your inability to work because of injuries received in an accident. This may be a very helpful type of coverage to include on your policy.

Loan or lease gap insurance can be purchased if you have a loan on the vehicle and have comprehensive and collision insurance with a deductible. This insurance covers you for the difference in the value of your car if it is damaged and the amount you owe on it at the time of the loss.

Other coverages that many people consider include a rental car reimbursement clause which covers the cost of renting a car while yours gets repaired. Towing and labor insurance is often referred to as roadside service and can cover the cost of towing your vehicle from the site of the accident to a repair shop.

If you have specialized custom parts or equipment, you can have those separately insured. This would include things such as a customized stereo, running boards, trailer hitches, customized wheels and spoilers, etc.

As you can see, there are many options for car insurance beyond the basics, and you may want to seriously consider whether or not they would be beneficial to you.

Just about every able bodied adult drives at least occasionally. And for those of us who own a car we understand how important it is to have your car covered in case of an accident. What most people do not realize is that in many states is the law that you have at least minimum coverage on your car for you to operate it on the road.

But who wants to pay tons of money for car insurance, especially with the state of the economy. Finding a good agent will help you keep your costs down while keeping your vehicle covered.

To know if you have the right amount of coverage for your vehicle you need to look at not just your state requirements but the requirements of your lean holder. That is right if you are making payments on your vehicle your lender has the right to require that you keep full coverage on it at all times.

Failure to do so can result in a breach of contract. That is of course if your lender spells out that you are to keep full coverage on it at all times. It is as much for their protection as it is for yours. You would not want to get into an accident, total your automobile and then be on the hook for those payments as well as new payments.

By requiring you to have full coverage your lender will be able to recoup the loss should something happen to your automobile.

But having full coverage automobile insurance just does not make a lot of sense to some if the vehicle is paid off, especially older cars. At some point in the cost of the premiums end up costing you more than the automobile is worth. For most states you can get away with having what is called liability insurance.

This means that your vehicle and you are not covered if you are found at fault in an accident. The others involved in the wreck will be covered up to a certain amount. Now the liability amounts vary from state to state so be sure to talk to your local agent before purchasing your policy.

When you are shopping for automobile insurance it is important to have an idea of what you need before you call. By taking your time and shopping around you will find that the price for car insurance will vary from company to company as will coverage.

By doing a simple online search you can get instant quotes from many of the companies in your area. By comparing quotes and coverage you will ensure that you are not over paying. And it really does come down to saving money and keeping yourself on the road.

CAR BASICS –Warranty!

A warranty is a promise made by the seller to the buyer regarding the quality of the product. If the product does not function properly, then the seller is liable to repair or replace the product, without any expense to the customer. It also specifies a certain time limit to redeem the warranty.

Type Of Warranty

A.Extended Car Warranty

Extended warranty is an additional warranty on a product after the actual warranty has expired. It can be purchased from the dealer of the product or even as an afterthought.

Whenever a consumer buys a product, he assumes that the product is fit to use. He trusts the features and functions as described or demonstrated by the seller, depending upon the type of the product. The buyer offers a warranty, promising to either fix or replace a faulty product.

Extended warranties can cost up to fifty percent of the purchase price. This again depends on the type of the product.

This warranty is only for a specific limit of time from the date of purchase. Cars often come with a "3 years or 36,000 miles" warranty. Other consumer durables too generally do not have a warranty of more than a year.

Many companies offer extended warranties on cars. In case of breakdowns and heavy repairs, they come in handy. Few companies even offer coverage against wear and tear of the parts of the car. However, different companies have different clauses in their warranties. Some companies prefer to exclude this coverage.

Dealers sell the extended warranty as insurance policy, which means they will replace the product in case of a breakdown after the standard warranty expires. In cases where wear and tear is not covered, this may not be worth the extra cost. The company may simply not pay the claim, citing the reason of breakdown as improper maintenance or regular wear.

Quotes for purchasing extended warranty can be found on the Internet on the websites of many leading companies. Some of them are warrantydirect.com, carbuyingtips.com and autowarranties.com

B. Implied Warranty

Implied warranties simply mean that the moment a sale is made, the seller automatically makes the promise that the product is in proper condition and there are no known defects present in it. Also, if later it is found that there are any problems, it is the manufacturers duty to fix or replace the product.

If the seller also makes any specific promise to the customer related to the product, and the product fails to deliver, it is still considered a breach of promise.

These warranties are assurances of the quality of the product at the time of sale and not how long it will last. Also, there are limitations on what can be claimed under the warranty. Any kind of mishandling or abuse of the product is generally not covered. Also, wear and tear due to regular use of the product may also not be covered under a warranty. This again could be product specific.

Sometimes, a warranty is applicable to used products, as well -- depending upon its type and price. If a seller does not want to offer any warranties, he has to notify the customer in advance. This is known as selling the product "as is".Of course, this may deter a customer from buying the product.

The seller may voluntarily offer an express warranty to a customer. This may be a part of a promotion or advertisement campaign.

CAR BASICS - Buying From A Car Dealer!

Buying a car is an exciting experience, but sometimes it is hard not to let that excitement and enthusiasm get the better of you. Automobiles are expensive, so you do not want to make any rash decisions.

Cars are no longer the luxury that only the rich and the famous could afford. Today, cars have become more of a necessity to every home and family. It is one of the most used means of transportation in the entire world. Therefore, you need to follow some new car buying tips before owing one of these beautiful pieces of machinery.

Today, there is a lot of competition, as far as car manufacturing is concerned. Different car manufacturers are consistently in the process of improving car designs and engine performance to attract more and more customers. In the end, it is the consumer, which is you, who gets to benefit in the bargain. You can buy one of the cool-looking cars at the best price if you do your homework well.

Things are better now for anyone who has not bought a car yet especially for most young people of today; it is one of the joys of a lifetime to find a car of your choice and buy it. Teens tend to purchase an automobile around the age of 17 or 18 which is the best time to start driving, and living out their young lives. The trouble is they never had much expertise in this field because of the lack of understanding the gimmicks behind buying vehicles.

Do your parents teach you about the problems when you go out researching? Of course, but that is not always the case. When young people get older they get wiser and feel that they can do things on their own without fault. Most of the time it is not their fault. What you need to know is that many new and used car dealerships prey on young people who do not know a lot about acquiring an auto.

Remember, it is not only young people that are affected by bad decisions when purchasing a vehicle.

Older people are victims of this nature also. Have all your paper work and possibly the bluebook for an auto purchase. Never leave home without it because it can determine how much is the exact amount on each vehicle that you come upon.

Sometimes dealerships may have cars that are a little over-priced which is not hard to see or understand due to the fact most of these vehicles are shipped directly to the dealer. When the dealer has the vehicle in his or possession then they randomly put a price on them. I do not care how much time you spend looking over the automobile, remember to ask tons questions before buying the car.

Below are some of the highlights and details of what you should know when buying a car or vehicle.Yes, you need to follow some useful tips to ensure that buying a new car does not turn into a stressful experience for you.

Here are some things you should consider to make sure you make the right choice.

- ***How to Pick the Right Dealership***

Choosing a seller is probably the best place to start. It is advisable not just to walk into the first one you come across and buying from them. Try asking a friend or family member for recommendations. This is an excellent way to ensure that you do not accidentally repeat the same mistake someone else you know has already made.

The most important factor that comes into play when choosing a car dealer is the type of vehicle you wish to purchase. Decide what class interests you, the number of people you need to transport on a regular basis, the price you are willing to spend, and the type of model you want. If you are looking for something sporty, the chances are that you will want a different dealership than if you want an SUV.

Once you know the type of dealership, try out a couple of branches if it is convenient. See how you feel about the salespeople who work there. Ask yourself, are they aggressive and going for a hard sell or are they friendly and approachable? It is imperative to feel comfortable as well as have room to make the right decision.

- ***How To Negotiate The Best Price***

Start by choosing the salesperson you prefer to do business with. Talk to a couple of people in the car dealership before you decide. Ask a few questions to break the ice.

Conduct extensive research before sitting down with the salesperson at the car dealer. Get an idea of the average price for that vehicle model to know what the car is worth.

Wrap up any negotiation quickly. Talk for half an hour or so and once you have made your offer, stay strong. Make sure it is approximately 5 percent of the dealer's price to be fair. Justify your offer and be consistent.

- ***Top Tips and Tricks***

Always take a test drive. You might have your heart set on a brand of car that you feel different about after taking it out on the open road. Be open-minded. If it does not feel right, be ready to find a different one that does.

Avoid answering personal questions at the car dealer. This is the salesperson's way of tailoring their pitch to you, which can make it a little difficult for you to make a decision based on opinion.

Before driving your automobile away from the seller, inspect it thoroughly. Any minor details that you are unhappy with should be rectified immediately, not a few weeks later.

Buying a vehicle is a big commitment, so take your time to find the right seller before jumping into it. Do your research on the market price as well as how to negotiate to get what you want.

- **Others**

Spend ample time to select the car of your choice: Buy a new car when you are in a position to wait for the right choice. This is one of the important new car buying tips. Do not ever bring yourself to a situation where your old car is in a broken down state and you desperately need a new car. Such kind of a situation can only profit the dealer and certainly not you. Therefore, start looking for the car of your dreams when you still have the time.

Get the best car loan and look for best incentives from different auto dealers: Before, you decide on the color and model of the car, make sure you have ready money to buy the car. If you want to finance the car, you need to ensure you know about the interest rates, monthly installment, and the tenure of the car loan. Check with different credit unions, banks, and other financial institutions to make sure you get the best deal. Another of the new car buying tips is to check with various auto dealers about the incentives that they offer. Incentives such as cash back, special financing deals, and customer loyalty discounts are things to look out for. Persuade the auto dealer to give you as much incentive as possible on the new purchase.

Do proper research and do not develop any emotions about a new car before buying it: Do proper research on the internet to study different models of your choice. You should know about the features and specifications of the model that you like before walking into a car shop. This will give you a sense of confidence and help build an impression of the salesperson who knows he or she cannot fool you. Another of the important new car buying tips is not to get emotionally attached to a car unless you bring it home. By doing so, you make it obvious to the salesperson that you love this car. This allows him or her to seal the deal of his or her choice and not yours.

CAR BASICS - Most Common Problems!

If our cars are starting to display some problems, different thoughts come into our mind regarding what caused the problem, how much will it cost to have it fix, do you need to buy anything and many more. That is why it is good for you as a driver that you have some knowledge on the most common car problems today so that you know how much you will spend and what it needs to be fixed.

Flat Tire - It is the most common problem that every car and other vehicle drivers face. Though this is the most common car problem that people face, it is among the easiest to fix car problems around. Having a spare tire, tube and needed tools for the job is very crucial for fixing a flat tire.

Shocks - Another very common car problem that nearly all car drivers will face throughout their lifetime driving is when their shock starts to malfunction. If you have noticed that your ride with your car is starting to become bumpy recently, then it is time for you to have your car checked by a car trusted and expert mechanic. If the problem is not very severe then the mechanic can have return your car to running smoothly, but if the problem is quite big then you might have to buy new shocks and shock absorbers and have the old ones replaced by it.

Will not Start Ignition - Just like what you see on movies, ignition is another common problem. The car will usually just crank but will not start; this is a common problem among cars and other vehicles. The problem may easily be fixed within just a few minutes or it can take days and a couple of cash if the problem is quite big. The cause for this problem is quite broad.

Cracked Windshield - The windshield is another very fragile and important part of your car. If you have a cracked windshield then it would be best that you fix it right away. Not fixing it by the time you noticed the crack will put you and your family in danger.

Having some basic knowledge on the basic problems and how to fix those car problems will help you know what the problem of your car is and what caused the problem. Aside from letting you have some knowledge about the problem of your car and what caused the problem, it will also help you keep you and your family safe from harm and it will also help you save money if you are able to prevent the problem from worsening.

CAR BASICS - Essential Auto Tools!

They say "better be safe than sorry"; the question is how to be safe? Well, to be prepared is the answer. In this world full of uncertainties, you never can tell when you may face a demanding situation. When it comes to automobiles, anything can go wrong without notice; this is why it is extremely important to carry all necessary auto tools with the automobile wherever you take it to, and also to be enough educated to perform critical repairs.

There are different types of auto tools required to perform different types of jobs. For motorcycle repair jobs you will need motorcycle tools, whereas for car repair jobs you will need car tools. When making the toolkit for your automobile, it is advisable to make a list of the tools you will need to perform the jobs you are skilled to perform. The ideal way to make the tool list is to inspect your automobile and look for the parts where a problem may arise and the tools you will need to fix it.

Some of the essential hand tools include emergency puncture repair kit, car jack, pliers, screw drivers, wrenches, air pump, and hammer. Other things you should keep are fuses, couple of headlight bulbs, wiring tape and some wires for fixing wiring-related issues. If you do not have enough information about how to execute repair jobs, you may use any video streaming website such as YouTube.com to search for repair job lessons as per the need. Some the most critical jobs you should be able to perform include bulb replacement, tyre replacement, and fuse replacement.

If you are planning to shop for car or motorcycle tools, you may use any search engine to find a list of top online tools stores offering an exhaustive range of automotive tools. You can browse their detailed catalogues to shop for the tools and supplies on the list. Shopping from an online store not only gives you a hassle-free shopping experience; the low overhead cost of maintaining an online store enables their owners to price the tools very reasonably.

CAR BASICS -Tools Should You Keep in Your Boot!

It is very important to watch the pennies in today's uncertain times. Repairing your car yourself is a good way to save money. If it is a minor problem then you should be able to fix it yourself providing you have the right car tools. Car mechanics do not have to be scary or worrying, with the right mindset you should be able to see it as an opportunity to broaden your horizons and learn something new. The market has an assortment of riches when it comes to car tools. You only need the necessary ones though and you should also make sure they are good quality.

There are many things that should, and need, to be kept in every motorist's car. Fluids and oils need to be replaced; hoses, lines, clamps, and filters are all subject to wear and tear. Roadside maintenance is nearly impossible without a basic set of car tools, but it does not have to be a daunting task if the driver is prepared. So, which car tools should every motorist keep in his or her boot?

Firstly, the hydraulic jack is a tool that you must have. This ensures that you will have easy access to the undercarriage of the car. This car tool will ensure your repairs, whether minor or major, go smoothly and comfortably. You should spend a little bit extra to make sure you have a quality jack as you do not want your car falling on your head.

Other car tools that you must have include screwdrivers, pliers, tyre wrench, ratchet and socket sets. As makes and models of cars are all different you need to bear in mind that you will need specifically sized car tools to accommodate this.

Pliers are versatile tools that will allow you to bend, cut and hold pieces of metal. There is a number of different pliers available which can be used for different purposes, these include locking pliers and groove joint pliers. Make sure you are using pliers that are suitable for the purpose.

Another thing you should always carry as part of your car tools is a set of **jump leads.** This will be of much convenience should your battery run low at any time.

One of the main factors to consider when buying car tools is how good their grip is. The more comfortable your car tools are to use, the easier you will find the job.

Invest in a decent **screwdriver**. There are many types of heads, and it may seem like this requires many individual tools, but luckily it is easy to find a quality screwdriver that has interchangeable heads. Often, a single tool will hold multiple bits. Two-sided bits, in particular, can offer the user two options for both flat and Phillips heads screws, the most common styles. It may also be wise to invest in a driver designed for the occasional Torx screw, or star shape.

Beyond screws, a roadside repair job will often have the user running into many types of bolts. To be prepared for this occasion it is advisable to have a **set of wrenches and socket wrenches** to hand. A basic set of tools can be easily purchased, to fit this need.

The tyre wrench will come in handy when removing nuts. You should make sure the tyre wrench ably fits the hexagonal nuts. Most manufacturers sell cars with a wrench that fits the nuts in the tyre. If you do not use the right size of wrench then you could end up damaging the nuts which may mean they are more difficult to take off.

Boxes of tools need not be huge, and for the average consumer it is possible to find a set that fits his or her needs.

There are a few things to consider when picking out a case of car tools. The buyer should look to see if there are both metric and U.S. standard units represented. It is common to run across a mixture of these standards of measure when repairing a vehicle. **Socket** sets should also come with at least a few extensions, to let the user reach around tight spaces.Many sets come with a few screwdriver heads and an assortment of hex wrenches, which are a great addition to any collection. The ratchet and socket sets will assist you in turning the bolts and nuts. They are available in a multitude of different sizes. However it is not advisable to use a socket wrench for spark plugs.This reduces the possibility of the spark plug cracking or shocking.

Beyond tools, there are a few other noteworthy items that every boot should not be without. *Electrical wire, electrical tape, duct tape, thin rubber sheeting and rubber hoses, motor oil, transmission and brake fluid, and a couple jugs of water* can all become necessary at one time or another. Standardized fluids are largely self-explanatory, but an odd collection of tape and spare parts may well be all a motorist needs to fix a leak and get the car to a repair station. It does not take long to prepare and the time spent will certainly be worth it when the car tools are needed.

CAR BASICS - Servicing!

Car servicing is something that every auto owner has to indulge in at least some time of the year. The best method to service your car is to go to your car service station and have the servicing done methodically; according to your car maker. Consider this. You car is up and running smoothly but the mileage figures show that you need to have it serviced. Certainly, this is the right thing to do but have you wondered what actually goes in to make you pay that servicing bill every 6 months?

Servicing your car is a simple DIY job that you can learn quickly and easily save some good money.

If you have never serviced your car before personally, you will get to know some of the servicing basics here in this chapter.

Car servicing essentially means inspecting the car thoroughly for any damaged components, and replacing some parts periodically that wear out over time. The parts that require inspection are the tire pressures, the brake fluid levels, air filters, oil filters, spark plugs, wheel balancing and alignment, battery fluid levels, etc.

If you review carefully it is certainly possible to do all this by yourself, as below-You will need a complete car tool kit as provided by your car maker when you purchased the vehicle. This will usually consist of all the spanners that you might need.

The correct grade engine oil. Engine oil needs to be replaced every 5000 miles (This figure will change as per the car type and make. Check your car manual for your car.) The type and grade of oil required for your car can be found in your car manual.

To get started, a typical car servicing will involve the following steps-

- ***Hosting your car up to check the underbody.***

This can be done on a ramp or using a jack. A quick inspection of the underbody will let you know if there is any damage.

- **_Replacing the engine oil_**

When you are done with this, you will now need to replace the engine oil. Always remember that the engine oil should never be checked or replaced while the engine is running. To proceed, you will need to unscrew a bolt that is just below the engine oil chamber and drain the oil in to a pan. When all the oil has been drained you will need to screw the nut back in tightly. You can also change the oil filter and it is usually just adjacent to the oil chamber. Use the correct type and size as specified for your car.

- **_You lower the car_**

You can lower the car now and fill the new engine oil. Be sure that you do not pour excess oil over the dipstick mark as this is going to be very difficult to drain off. Wait patiently for the oil levels to settle.

- **Check the spark plugs**

The other things you will need to check include the spark plugs. Spark plugs need to be cleaned periodically. Unscrew the spark plugs from the engine and clean them with a soft cloth. You will need to shine the plug contact points using sandpaper. Fit them back in, and now check out the air filter. Cleaning the air filter is easy and you can use your vacuum cleaner to do the job.

- **Checking the suspension**

You will also need to check out the suspension. Bump the car and observe how long it takes to settle. Ideally this will take no more than 3 bumps. Also check out for any spills, or leaks coming out of the shock absorbers.

- **Check the brake fluids level**

Now you will need to check out levels of brake fluids and top them up if necessary. Also, inspect the brake shoe condition by removing the wheel and inspecting the braking unit. If worn out, you can easily change them on your own.

With this done you are completed with the major servicing part of the car. The remaining things include oiling your doors and the hinges, checking if your lights are working, testing if your seat belts are working, and measuring the tire pressures (front and rear).

This might seem a long list to do but if you observe a typical car servicing process you will understand the nuances within no time. An easy way to proceed is by creating a check list for the different things will need to do. This way you can be sure of not missing out on any parts that need to be serviced.

Completely servicing your car will need a few hours but you will end up saving a lot of money. Besides, it is fun too and a lot of learning.

CAR BASICS - Most Common Road Traffic Offences!

Over the years, driving has turned out to be the real fun because vehicles have become safer, more eco-friendly and fuel efficient. At the same time, quality of driving is becoming poorer and police has to deal with the millions of traffic offences every year. Though, if you believe that you are a very good driver but still there are some set rules or laws that you should follow to assure a safe driving.

There are various types of traffic violations that you should avoid and becoming vigilant of these offences will assure that you are abiding by the law and will also help you from being prosecuted by the law. In this book, we will discuss about some of the most common road traffic offences that can put you in trouble. **Below are some of them;**

- *Neglecting Speed Limits*

Speeding is considered as the most common traffic offence. Most of the drivers drive above the speed limit in urban areas. If you are caught speeding you may get disqualified from driving or points on your license. You could also be charged with the fine or charge certain fees. The speeding limit for the urban areas is 30mph and 60mph on the carriage ways.

Using Mobile Phone While Driving

Avoid making calls or texting while driving because it is illegal to drive using mobiles or hand held phones, while this law does not implies directly on using hands free phone. You can only make an emergency call when it is not viable to stop.

Driving Under Alcohol or Drugs Influence

According to some reports, most of the accidents and violation cases occur when driver drives under the heavy intake of alcohol or drugs. To avoid any kind of offences, it is crucial to stay away from any kind of unlawful substances.

- **Dangerous Driving**

Traffic offences occur when driver uses unsafe techniques of driving resulting into damage of property or lives. It is considered as crime and is punishable by lengthy imprisonment. You can avoid this by driving on safe side of road, taking driving lessons and obeying traffic rules.

- **Driving with Suspended License or without License**

If you are driving without license, it is considered as crim equal to the theft or fraud. If your license has suspended or you don't have license and you need to go out in emergency, then it is better to take your friend to drive instead of driving yourself.

You should need to be aware of the traffic laws of your country, so that you will always remain free from the criminal charges that are implied due to road traffic offences. Be careful, be safe!

CAR BASICS - Accessories Are a Must For Every Car!

Everyone loves his or her car, as a car is one of the most costly items you will ever purchase in your life. People spend lots of money in making their car look good and this can be done by adding various car accessories. These accessories change overall look, style and grace of the car. These accessories are mainly divided into two categories- ***external accessories and internal accessories***, depending on where you want to use them. These accessories come in lots of design and style. You have to choose those accessories, which compliment your car's style. These accessories are add-ons for your car and they will definitely enhance your car's look.

If you search on internet or you visit any car accessory shop, you will definitely come across with lots of accessories for your car. For external accessories, you will get things like ***spoilers, car cover, fog lamp, wheel cover, alloy wheels etc, and for internal accessories, you will get MP3 players,***

speakers, woofers, seat covers, dashboard cover, air perfume, variety of floor mats etc. You can spend as much money on these accessories because this market has endless variety.

First accessory for your car is its **music player and sound system.** Every music lover would love to have best quality of MP3 player for his car. You can also add video player with your MP3 player and by doing this, you will be able to watch videos while driving. L.E.D headlight is also gaining popularity these days as it changes overall look of your car. These lights look stunning while driving and they will help you in foggy and dusty atmosphere. You can also change design and style of side mirrors by adding indicators on them. Seat covers are also necessary for your car as they play a vital role in over all look of car. There are various types of seat covers like cotton, leather etc. You just have to select according to your car's color and model. There are few more car accessories like sensors, emergency kits, jumper cables etc.

There is huge variety of car accessories available in market. These accessories play a vital role in making your car looks beautiful. You can change internal as well as external looks of your car by adding variety of accessories. Few of them are necessary while many other are just optional like spoiler, fog lamp, alloys etc. There is no limit of these accessories and you can spend thousands in buying them.

You can get these accessories from any renowned shop. You can also buy them online as lots of websites deal in these accessories. They will also offer you some discount if you buy from them. However, it is always advisable to buy them from any renowned shop or from wholesale market. Prepare a list of accessories you want to purchase, visit any renowned shop and then ask for latest design of accessories for your car

CAR BASICS– Specific Accessories For Specific Reason!

May be to do up a used car or to capacitate a brand new one, car accessories are necessary in both the cases. A car without proper accessories is like a house with no furniture or a garden with no flower. The home or the garden may be known by the name that their shape deserves, but they will not fully fit for human living or proving pleasure. Same is the case with a car that lacks in adequate accessories.

Basically accessories are necessary to equip a car for higher comfort and better capability. A car must have some of the most important accessories without which it will fail to give necessary services to its owner. Some accessories are there that one may do without, but very much important in beautifying a car. Without these the car may be able to serve the purpose of its owner; but it may not be attractive at all.

So, to make your car stand out from the rest and a comfortable place to stay in, adorning it with accessories is a must. Among the ornaments a car can be garnished with dash kits, body kits, alloy wheels, car security and alarms, lockwood dials rear spoilers, snooper DPS detectors, roof boxes, racks etc. are mostly used. Interior car styling accessories like gear knobs, tax disc holders, pedal sets, leather gaiters, handbrake handles and loads are frequently in use.

Winter car accessories like snow chains and frost protection tybond car covers, PIAA wiper blades, uprated headlight bulbs are among the seasonal equipments. Exterior body styling like side skirts, front DTM spoilers and light brows and masks are used to give the car a classy look. Use of various types of lights like rear lexus style lights, halo ring headlights, indicators and side repeaters add attraction to any car and contribute to increase its security level as well.

CONCLUSION

Cars are vital to most of our every day lives. They transport us to and from our homes, to school, our places of work, to our daily errand runs and activities and so much more, it would be hard to name them all. Cars allow us to do all of this in a blink of an eye, extremely fast and all with little to no trouble at all. They allow us to go very long and far distances in a snap, as many times as we desire and carry on with our lives.

So when you really think about it and everything that cars enable us to do on a daily basis, it really is a huge blessing to even have a car. That is why its important for us to perform basic car maintenance on a regular basis in order to ensure that our autos run at it's best with little to no interruption at all. Making sure that the car receives basic car maintenance will ensure that it runs smooth and will not stall or stop on you.

www.ingramcontent.com/pod-product-compliance
Lightning Source LLC
Chambersburg PA
CBHW071022240526
45469CB00006BD/2040